The Voyage of the *Beagle*

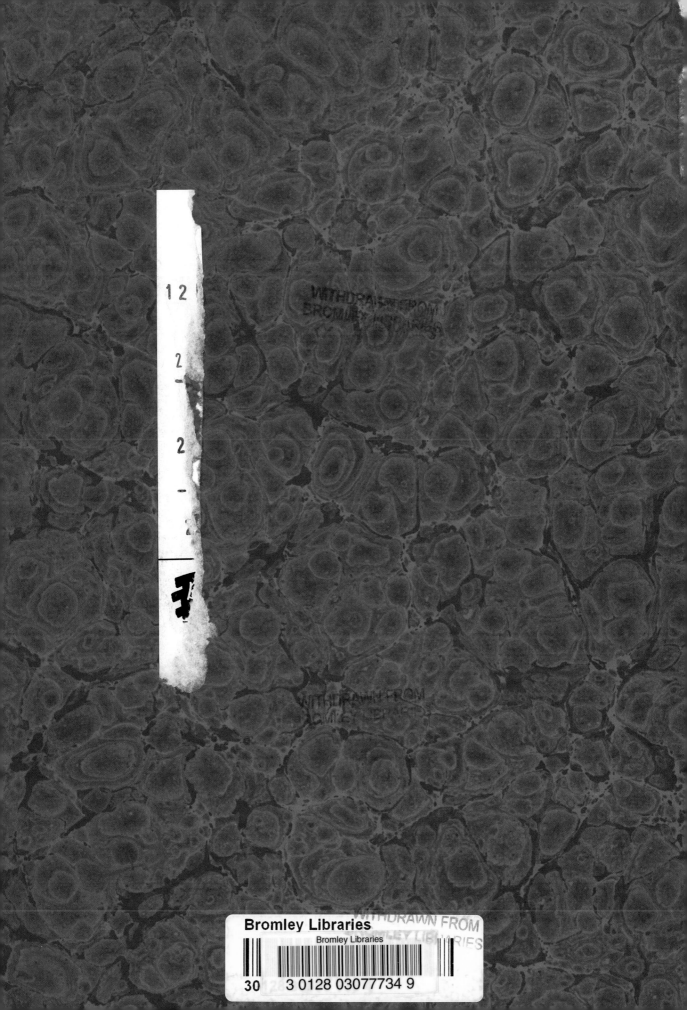

The Voyage of the
Beagle

Darwin's extraordinary adventure aboard FitzRoy's famous survey ship

James Taylor

CONWAY

For my parents – Stewart and Elizabeth Taylor

A Conway Maritime book

Copyright © James Taylor 2008

First published in Great Britain in 2008 by
Conway
an imprint of Anova Books Company Ltd
10 Southcombe Street
London W14 0RA
www.anovabooks.com

British Library Cataloguing in Publication Data
A record of this title is available on request from the British Library.

ISBN: 9781844860661

Edited by: Alison Moss
Designed by: Georgina Hewitt
Printed and bound by SNP Leefung, China

To receive regular email updates on forthcoming Conway titles, email conway@anovabooks.com
with Conway Update in the subject field.

Half title page: 'Monte Video, Mole' (detail), engraving after Augustus Earle from the *Narrative* (1839). Today Montevideo is the largest city, capital and chief port of Uruguay. The mole in Earle's composition refers to the pier that acts as a breakwater and also as a cargo-handling facility.

Frontispiece: Fuegians and the *Beagle* at Portrait Cove in Beagle Channel, Tierra del Fuego, watercolour by Conrad Martens. Robert FitzRoy discovered the Beagle Channel during the survey ship's first expedition to South American waters. He also selected this image as an engraved illustration for the *Narrative* (1839).

Opposite: Tanagra darwinii, coloured lithograph from *Birds* Part 3 No. 4 of *The Zoology of the Voyage of H.M.S. Beagle* by John Gould. Several species brought back from the *Beagle* expedition were named after Darwin, their discoverer.

Contents

TOWN SEEN ACROSS THE RIVER.

OLD LOOK-OUT TOWER.

VALDIVIA.

Author Biography

James Taylor MA (Hons) FRSA studied at the Universities of St Andrews and Manchester. Formerly Head of Victorian Paintings at Phillips Fine Art Auctioneers, from 1989 to 1999 he was a curator of paintings, drawings and prints, exhibition organizer and corporate membership manager at the National Maritime Museum, Greenwich. He contributed to and co-organized the *Polar Exploration* galleries, *SeaPower* galleries, and the exhibitions *Henry VIII: A European Court in England* and *Titanic*. From 1999 he has been a freelance art consultant and exhibition organizer, and in 2001 he was appointed a NADFAS lecturer. Publications include illustrated histories: *Marine Painting* (Studio Editions, 1995) and *Yachts on Canvas* (Conway, 1998), and the catalogue entries for the 2007 exhibition *Rule Britannia! Art, Power and Royalty* to mark the 400th anniversary of Jamestown. Received the Sir James Caird and Sir Geoffrey Callender awards for writing and supporting the National Maritime Museum's public lecture programme. His lectures for the National Association of Decorative and Fine Art Societies (NADFAS) include 'Captain James Cook and the Art of Exploration' and 'Charles Darwin and the Voyage of the *Beagle*'.

Opposite: 'Valdivia', engraving after Phillip Parker King from the *Narrative* (1839). Valdivia is Chile's 'city of rivers' and today is home to around 140,000 people.

Overleaf: Tahiti (detail), engraving after Conrad Martens from the *Narrative* (1839).

Acknowledgements

Two internet resources: Charles Darwin Online www. darwin-online.org.uk and the Darwin Correspondence Project www.darwinproject.ac.uk (featuring more than 14,000 letters) have made the process of searching and examining illustrations, letters, manuscripts, publications and associated material considerably easier for everyone. Their staff and supporters deserve universal praise.

The Darwin Online project, led by Dr John van Wyhe, obtained permission from English Heritage to reproduce Darwin's original version of his *Beagle* Diary ('Journal'), which he sometimes referred to as his Logbook. It reveals many important nautical and shipboard references that were omitted from previously published and heavily edited versions. This Diary is kept at Down House.

Jamie Owen of The Royal Geographical Society, and the library staff, were supportive from the outset. Likewise Liza Verity, Amy Miller and the librarians of the National Maritime Museum, Greenwich, provided invaluable assistance. Thanks also to Javis Gurr and Jonathan Butler, English Heritage; Paul Sturm and Derek Clear, The National Archives, Kew; Guy Hannaford and Liz Hill, British Hydrographic Office; Christine Woollett, The Royal Society; Adam Perkins, Cambridge University Library; Candace Guite and Ann Keith, Christ's College, Cambridge; David McClay, John Murray Archive; Bernard Horrocks, National Portrait Gallery, London; Anna Smith, The Wellcome Library; Louise King, Royal College of Surgeons; Lynn Miller, The Wedgwood Museum and Wedgwood Museum Trust; Theresa Calver, Colchester and Ipswich Museum Service; David Taylor, Colin Starkey, Doug McCarthy, Gudrun Muller of the Picture Library, National Maritime Museum, Greenwich; Ruth Long and Paul Scudder, Imaging Services, Cambridge University Library; Norbert Ludwig, Picture Library, State Museums, Berlin; Nicholas Schmidt, Pictures Branch, National Library of Australia, Canberra; and Jennifer Broomhead, Intellectual Property and Copyright Librarian, State Library, New South Wales.

I had the good fortune of meeting, by chance, Mary Eagle in the Caird Library, National Maritime Museum, and exchanged information on Augustus Earle. Mary clarified some key points of Earle's life and career, and indicated the whereabouts of a portrait of John Clements Wickham, FitzRoy's first lieutenant. In return I was pleased to flag up the remarkable letter dated 4 January 1818 from Augustus Earle to Sir Joseph Banks in the National Archives, Kew (see Chapter 6).

Elizabeth Ellis, Curator of Pictures, Mitchell Library, State Library of New South Wales, and specialist on the life and work of Conrad Martens (who identified Conrad Marten's shipboard *Journal*) kindly read through the manuscript in draft form and offered several pointers and leads to other sources.

Thanks to John Lee of Conway for having faith in the project, and Alison Moss, my editor for keeping things in order, Matthew Jones for co-ordinating the acquisition of picture images, and Georgina Hewitt for diligently designing this publication.

I would especially like to thank Dr Gordon Chancellor, Associate Editor of the Charles Darwin Online project, for his valuable insights and for sharing with me his *Beagle* enthusiasm.

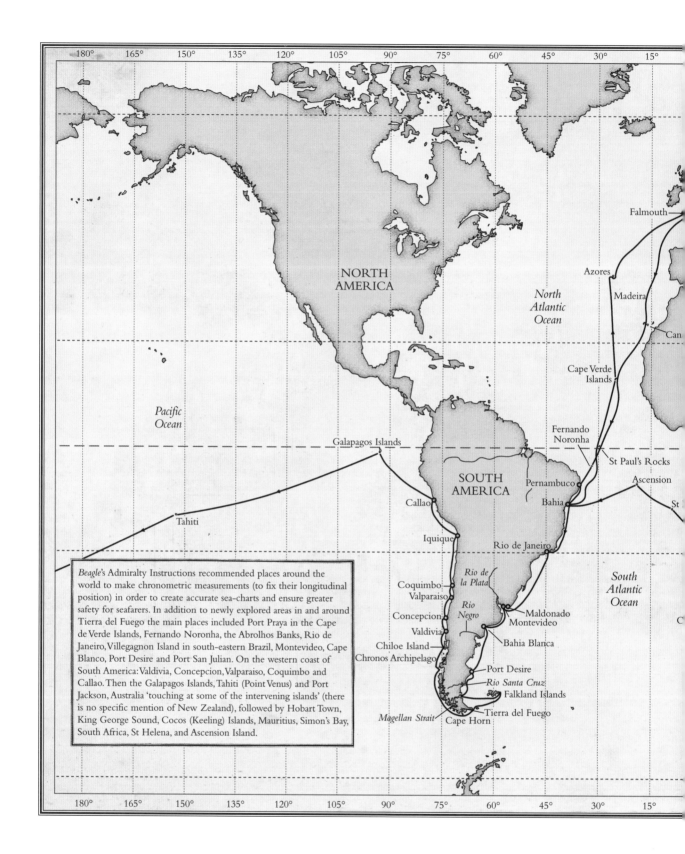

Longitude labels (top): 180° 165° 150° 135° 120° 105° 90° 75° 60° 45° 30° 15°

NORTH AMERICA

Pacific Ocean

North Atlantic Ocean

Falmouth

Azores

Madeira

Can

Cape Verde Islands

Galapagos Islands

Fernando Noronha

St Paul's Rocks

SOUTH AMERICA

Pernambuco

Ascension

Callao

Bahia

St

Tahiti

Iquique

Rio de Janeiro

South Atlantic Ocean

Coquimbo

Rio de la Plata

Valparaiso

Rio Negro

Concepcion

Maldonado

Montevideo

Valdivia

Bahia Blanca

Chiloe Island

Chronos Archipelago

Port Desire

Rio Santa Cruz

Falkland Islands

Magellan Strait

Tierra del Fuego

Cape Horn

Beagle's Admiralty Instructions recommended places around the
world to make chronometric measurements (to fix their longitudinal
position) in order to create accurate sea-charts and ensure greater
safety for seafarers. In addition to newly explored areas in and around
Tierra del Fuego the main places included Port Praya in the Cape
de Verde Islands, Fernando Noronha, the Abrolhos Banks, Rio de
Janeiro, Villegagnon Island in south-eastern Brazil, Montevideo, Cape
Blanco, Port Desire and Port San Julian. On the western coast of
South America: Valdivia, Concepcion, Valparaiso, Coquimbo and
Callao. Then the Galapagos Islands, Tahiti (Point Venus) and Port
Jackson, Australia 'touching at some of the intervening islands' (there
is no specific mention of New Zealand), followed by Hobart Town,
King George Sound, Cocos (Keeling) Islands, Mauritius, Simon's Bay,
South Africa, St Helena, and Ascension Island.

Longitude labels (bottom): 180° 165° 150° 135° 120° 105° 90° 75° 60° 45° 30° 15°

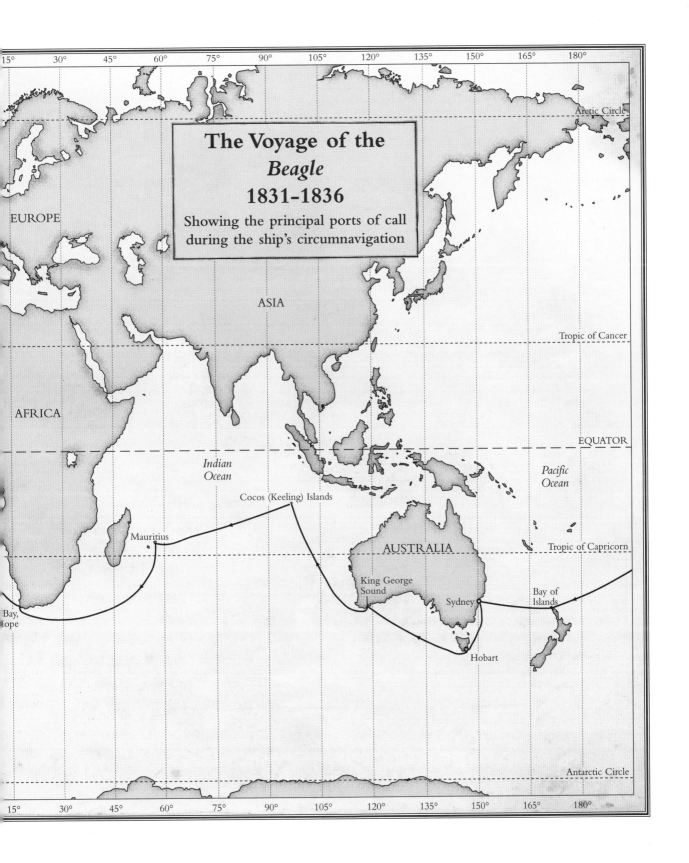

The Voyage of the
Beagle
1831–1836

Showing the principal ports of call
during the ship's circumnavigation

Introduction

Charles Darwin, albumen print (1868) by Julia Margaret Cameron. Darwin regarded this profile photograph of him (one of three she took), looking serene and thoughtful, as the finest portrait of all. He paid her £4 7s, and additional sums for other photographs.

Millions of people around the world have touched a portrait of Charles Darwin. He appears on the back of the English £10 note. His image derives from a print by the pioneering Victorian photographer Julia Margaret Cameron and the design went into mass circulation in the autumn of 2000. In addition to Darwin's bearded profile, the note features his personal magnifying glass, stylized flora and fauna, as well as a depiction of HMS *Beagle*.

The voyage of the *Beagle* has become synonymous with Darwin's name, although in recent years a small number of biographies have helped to redress the balance by focusing on the significant role of Robert FitzRoy, the ship's commander.

Robert FitzRoy should not be described simply as 'Darwin's captain'. He was a remarkable seafarer and nautical surveyor. In 1854 FitzRoy was selected to head up a new experimental government department within the Board of Trade of what is now known as the Met Office. He became the world's first full-time weather forecaster, and in fact invented the term 'weather forecast'. Today the headquarters of the Met Office in Exeter is situated on FitzRoy Road. FitzRoy's concern was for the safety of all sailors at sea and he also contributed greatly to the work of the Royal National Lifeboat Institution (RNLI). In February 2002 the sea area Finisterre off the north-western coast of Spain was renamed FitzRoy in recognition of his services.

Without FitzRoy it is unlikely that Darwin would be the household name of today. He would have followed his chosen career as a country clergyman and pursued science as a hobby rather than a profession. FitzRoy deserves greater recognition for appointing Darwin to the ship, enabling him to observe, study and collect. He encouraged him to write up his notes in a methodical manner.

FitzRoy invited Darwin to contribute to the substantial publication that will hereafter be referred to as the *Narrative*: the full title being the rather long and ponderous *Narrative of the Surveying Voyages of His Majesty's Ships Adventure and Beagle Between the Years 1826 and 1836: Describing Their Examination of the Southern Shores of South America, and the Beagle's Circumnavigation of the Globe*. It was published in four volumes by Henry Colburn, London, in 1839. FitzRoy compiled and edited the first and wrote the second of the main volumes (the fourth was essentially a lengthy appendix). Darwin produced the third, which bears the subsidiary title *Journal and Remarks*, that would later be published separately to FitzRoy's volumes, also in 1839. Since then the name has changed several times and the text has undergone significant revisions. Known today as *The Voyage of the Beagle*, it is still in print and ranks as one of the world's bestselling travel books.

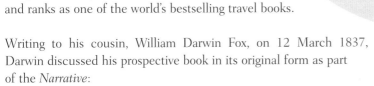

Writing to his cousin, William Darwin Fox, on 12 March 1837, Darwin discussed his prospective book in its original form as part of the *Narrative*:

> *In your last letter you urge me, to get ready the book. I am now hard at work and give up every thing else for it. Our plan is as follows.— Capt. FitzRoy writes two volumes, out of the materials collected during both the last voyage under Capt. King to T. del Fuego and during our circumnavigation.— I am to have the third volume, in which I intend giving a kind of journal of a naturalist, not following however always the order of time, but rather the order of position.— The habits of animals will occupy a large portion, sketches of the geology, the appearance of the country, and personal details will make the hodge-podge complete...*

Robert FitzRoy, albumen print (1850s) by the London Stereoscopic & Photographic Company (active 1854–1908). This highly successful company specialized in all areas of photography, from retail and publishing work through to customized studio portraiture. It was also a leading supplier of cameras and photographic equipment.

Darwin developed his volume (the 'Journal') from his detailed shipboard Diary. Writing to family and friends from the *Beagle* he would often refer to his Diary as his Journal and sometimes his Logbook. This has created unintentional confusion but it has been confirmed that the Diary was produced first. The original Diary containing 770 pages of writing is now the property of English Heritage and is kept at Down House. It has recently become available via the Darwin-Online website in its full unedited form.

NARRATIVE

OF THE

SURVEYING VOYAGES

OF HIS MAJESTY'S SHIPS

ADVENTURE AND BEAGLE,

BETWEEN

THE YEARS 1826 AND 1836,

DESCRIBING THEIR

EXAMINATION OF THE SOUTHERN SHORES

OF

SOUTH AMERICA,

AND

THE BEAGLE'S CIRCUMNAVIGATION OF THE GLOBE.

IN THREE VOLUMES.
VOL. I.

LONDON:
HENRY COLBURN, GREAT MARLBOROUGH STREET.

1839.

Title page from Volume I of the *Narrative*, published by Henry Colburn in 1839. Due to popular demand (and an agreement Darwin had made with FitzRoy) Darwin's contribution (Volume III) was published separately later in the same year. Its title has changed several times and is known today as *The Voyage of the Beagle*.

In 2005, during a year of celebrations to commemorate the bicentenary of the death of Vice-Admiral Horatio Nelson, there were also a small number of low-key events to mark the bicentenary of the birth of Vice-Admiral Robert FitzRoy. Both men were motivated by loyalty to their country and driven by a devout sense of duty. Although the two men were very different personalities they shared many qualities in common: they were inspirational and charismatic leaders but petulant and over-sensitive to criticism. In terms of discipline they were firm but fair, and they earned the respect and loyalty of their men. They acted independently and disobeyed official orders and instructions to achieve their goals; and did not expect any man to do anything they would not do themselves. However, in middle age Nelson would die in battle with honour. Around retirement age FitzRoy tragically would take his own life.

FitzRoy had carefully selected Darwin. He wanted a naturalist who would also be his messmate and companion, and so he had insisted that this position should be offered only to someone of a comparable social class – a gentleman.

Beagle's captain was conscious of a history of depression and suicide within his own family. Darwin confirmed this fact in a letter to his sister Catherine from Valparaiso (8 November 1834), writing that FitzRoy was 'aware of his hereditary predisposition'. FitzRoy's uncle, the British statesman 'Lord Castlereagh', had killed himself in 1822.

So it is likely that as part of FitzRoy's detailed planning for *Beagle*'s survey expedition he saw in Darwin someone with whom he could confide. FitzRoy had interviewed and appraised his character in London. Darwin was sailing in a private capacity and was not subject to the usual naval chain of command and discipline. He was asked by FitzRoy to dine in the captain's cabin three times a day, for almost every day, while at sea.

Wittingly (or unwittingly) this turned out to be prudent planning on FitzRoy's part, because during the voyage his senior officers, and Darwin, were called upon to help him through a period of depression. If FitzRoy had resigned as captain (as he threatened to do) the *Beagle* expedition would have been prematurely terminated.

During the *Beagle* voyage Darwin and FitzRoy became friends. Darwin respected FitzRoy's strong moral stance, and weathered the quarrels that occurred onboard due to the captain's hot temper. Politically they were poles apart. Darwin was a liberal Whig who opposed slavery and FitzRoy was a staunch Tory who believed it was a necessary evil. FitzRoy had been prewarned of Darwin's political leaning before his appointment. It is still surprising that there were relatively few disputes between them.

Although Darwin lacked any formal scientific qualifications he proved to be a brilliant naturalist. He had been tutored by some of the finest scientists in Britain and possessed just the right kind of youthful enquiring mind that could make connections, ranging across thousands of miles, between the flora and fauna, as well as the people and places, encountered on the expedition. He was a man who was capable of seeing both the wood and the trees. Darwin's wide-eyed wonderings, observations, measurements and prolific note-taking during the *Beagle* expedition would later be worked up into a series of ground-breaking publications.

In addition to the considerable monies paid out by the British Admiralty to alter and rebuild significant parts of the *Beagle* for her survey voyage, FitzRoy also spent a substantial personal fortune on the expedition, and thanks to him the *Beagle* was the finest equipped survey ship of the early nineteenth century.

The Admiralty's selection of HMS *Beagle* was a curious one. She was of a class of vessel with a very poor seafaring reputation and an abysmal safety record. No doubt the alterations and refit helped to make her more seaworthy; however, we should not underestimate the professional seamanship of FitzRoy, his officers and crew, who ensured that the ship returned safely to England after a voyage lasting more than four and a half years.

The voyage of the *Beagle* made famous by Darwin was actually the second expedition the ship made to South American waters. The main reason for the first expedition was all bound up in the changing history of South American politics. The end of the Napoleonic Wars (1803–1815) ushered in a protracted period of peace, known as *Pax Britannica*, during which Britain could focus on exploration, long-distance trading, and empire building. Several Spanish colonies in South America, including Argentina, Chile and Peru, had recently fought and won freedom during the wars of independence, 1814–24. Spain, Portugal, Britain, France, North America and, to some extent, Russia were all now competing for influence and trade. To this end Britain established and maintained naval stations to further diplomatic relations in Argentina, Brazil and Chile, and other areas of South America. In fact, by the time *Beagle* arrived in South American waters on her second expedition, British speculative investments across the continent in mining and railroads, among many other businesses, was already considerable, estimated at tens of millions of pounds. But to achieve effective trade, reliable knowledge of coastlines and navigable passages was required. Almost all of the existing charts and maps were known to be inaccurate.

The British Hydrographic Office, established in the late eighteenth century to create coastal charts ('maps of the sea'), was tasked to supervise the survey work, and in 1825 the Lords Commissioners of the Admiralty ordered that two ships be prepared for a survey of the southern part of South America. On 22 May 1826 HMS *Beagle* set sail from Plymouth with Commander Pringle Stokes as her first captain, but under the overall

Overleaf: 'Part of Tierra Del Fuego from H.M.S. Beagle 1834', from the *Narrative* (1839), published by Henry Colburn and engraved by J. & C. Walker. The Walkers were London-based engravers draughtsmen and publishers, who produced a large number of British Admiralty charts.

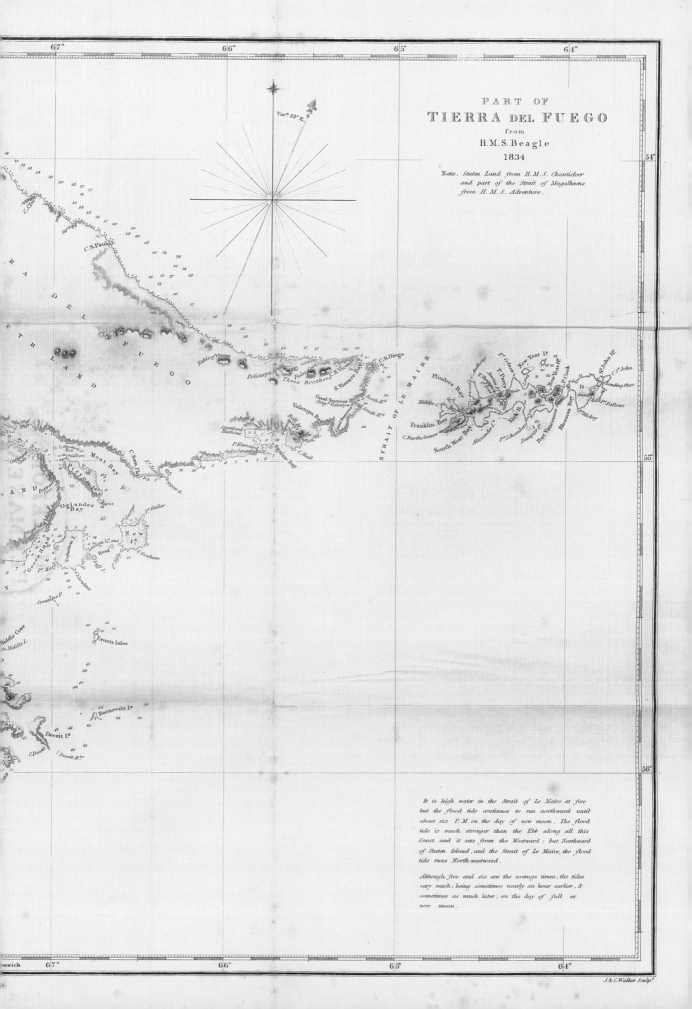

PART OF
TIERRA DEL FUEGO
from
H.M.S. Beagle
1834

Note. Staten Land from H.M.S. Chanticleer
and part of the Strait of Magalhaens
from H.M.S. Adventure.

Var.ⁿ 23° E.

It is high water in the Strait of Le Maire at five
but the flood tide continues to run northward until
about six P.M. on the day of new moon. The flood
tide is much stronger than the Ebb along all this
Coast and it sets from the Westward : but Northward
of Staten Island, and the Strait of Le Maire, the flood
tide runs North-westward.

Although five and six are the average times ; the tides
vary much ; being sometimes nearly an hour earlier, &
sometimes as much later ; on the day of full or
new moon.

J. & C. Walker Sculp.ᵗ

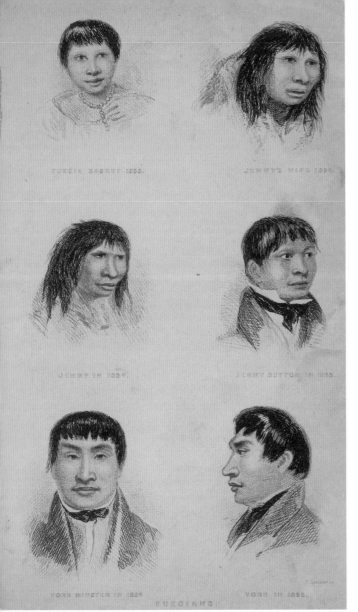

FitzRoy's Fuegians, engravings from the *Narrative* (1839) after drawings by Robert FitzRoy. Surprisingly these are the only images of York Minster, Jemmy Button and Fuegia Basket extant. Presumably FitzRoy prevented artists from gaining access to them.

command of Captain Phillip Parker King, who commanded HMS *Adventure*, a larger transport brig serving as a support vessel. King had already acquired a formidable reputation as a naval hydrographer through creating charts of large parts of western coasts of Australia.

King acted upon the following Admiralty orders:
Whereas we think that an accurate Survey should be made of the Southern Coasts of the Peninsula of South America, from the southern entrance to the River Plata, round to Chiloe, and of Tierra del Fuego, and whereas we have been induced to repose confidence in you, from your conduct of the Surveys in New Holland [Australia]; we have placed you in command of His Majesty's Surveying Vessel the Adventure; we have directed Captain Stokes, of His Majesty's Surveying Vessel the Beagle, to follow your orders.

The orders continued:
Both these vessels are provided with all the means which are necessary for the complete execution of the object above-mentioned, and for the health and comfort of the Ships's Companies. You are also furnished with all the information, we at present possess, of the ports which you are to survey; and nine Government Chronometers have been embarked in the Adventure, and three in the Beagle, for the better determination of the Longitudes.

Stokes was gradually overwhelmed by the arduous survey work and atrocious weather conditions. In May 1828 he reached his lowest point when he recorded in his Journal: 'The weather was that in which the soul of man dies in him'. At Port Famine in the Strait of Magellan, after a prolonged period of depression, he shot himself. Lieutenant William George Skyring took over temporary command of the vessel, although Admiral Robert Waller Otway, the commander-in-chief of the South American Station, decided to appoint one of his favourites, Robert FitzRoy, as captain of the ship for the rest of the voyage.

During FitzRoy's command a group of Fuegians stole one of the *Beagle*'s whaleboats. His abortive attempts to retrieve the craft resulted in him taking four of them onboard the *Beagle* as hostages and interpreters to help get his boat back. It was never retrieved and he controversially decided to take the Fuegians back to Britain as part of an experiment. They would be educated in England, converted to Christianity, and then returned to Tierra del Fuego, where they, in turn, would help to civilize their own people and be of practical assistance to visiting British seamen. That was the theory, but the reality turned out to be far from beneficial.

FitzRoy was initially promised a ship by the Admiralty to return his natives, but later the Admiralty reneged on the pledge, probably due to a combination of political and economic reasons, and so he began the process of funding his own expedition. Fortunately official Admiralty approval was eventually granted thanks to canvassing by FitzRoy's influential family members. Captain Francis Beaufort, the Admiralty Hydrographer, was convinced of the merits of a second expedition and he drew up extensive instructions of where and what should be surveyed. These were succinctly summed up in Darwin's Diary: 'The object of the expedition was to complete the survey of Patagonia and Tierra del Fuego (started on *Beagle*'s first voyage)…to survey the shores of Chile, Peru, and some of the islands of the Pacific; and to carry a chain of chronometrical measurements round the world'.

Stopovers in various countries around the world were required to complete the 'chain of chronometrical measurements'; or, in other words to accurately fix their longitude using the ship's chronometers. FitzRoy's success in completing this 'chain' during the *Beagle* voyage was cited as one of the supporting sections for his election to Fellowship of The Royal Society in 1851.

On the outward leg the *Beagle* explored part of the Cape Verde Islands off the western coast of Africa. Arriving in the north-eastern part of South America she then sailed south to Bahia (Salvador), followed by a prolonged residence at Rio de Janeiro. The survey party headed for Montevideo, Maldonado and Buenos Aires. She stopped at Port Desire and further south explored part of the Santa Cruz River. Two visits were made to the Falklands Islands.

The *Beagle* successfully made extensive surveys of Tierra del Fuego and the Strait of Magellan. She headed up the western coast of South America to explore the Chronos Archipelago, Chiloe, Valdivia, Concepcion and Valparaiso, with inland expeditions to Santiago, followed by Coquimbo and Copiapo. After stopovers at Iquique and Arica the ship's last port of call was Callao in Peru before setting sail for the Galapagos Islands. To complete the global chain of meridian measurements, the expedition called at Tahiti, the Bay of Islands in New Zealand, Sydney in Australia, Hobart in Tasmania, moving on to the south-western part of Australia for King George Sound. Heading west she examined the Cocos (Keeling) Islands and called at Mauritius. The final series of homeward-bound legs included the Cape of Good Hope, St Helena, Ascension Island and the Western Isles (Azores) before returning to England.

But by far the larger part of time was spent surveying in South America. This fact was summarized by Dr Janet Browne: 'Of the 57 months that the Beagle was at sea 42 were spent in the waters of South America. Of these, 27 were spent on the east coast and 15 on the west'.

So the *Beagle*, under the command of Robert FitzRoy, left Devonport, Plymouth, on 27 December 1831, ostensibly to complete the South American survey work and undertake other Admiralty instructions, but in reality FitzRoy had forced their hand and she was commissioned to return

Book spine of *On the Origin of Species* (1859). One of publisher John Murray's literary advisors, the Reverend Whitwell Elwin, was asked his opinion of the manuscript. Elwin advised against publishing the book, describing it as a 'wild and foolish piece of imagination'. The first printrun of just 1,250 copies, priced at 14 shillings, and bound in green publisher's cloth with no eye-catching illustrations, sold out to the book trade on the day of publication.

Pages from one of Darwin's notebooks (probably compiled in March 1835). They are full of detail and were used to write up his shipboard Diary, that in turn was used as source material for his volume of the *Narrative*.

This book examines the fascinating series of interconnecting stories relating to *Beagle*'s celebrated second survey expedition. The dominant influence of Sir Joseph Banks, President of The Royal Society, and his relationship to the British Admiralty and the Hydrographic Office is outlined. Also included are details of earlier voyages of discovery and the naval academies and schools that encouraged and taught the trainee naval officers to become proficient in drawing to aid their survey work. Chapters include the origin and design of the survey ship, the driving forces, characters and achievements of the ship's captain, officers and crew, the role and work of Charles Darwin, the inventive and often arduous nature of the survey work, as well as the remarkable creative work of Augustus Earle and Conrad Martens, FitzRoy's artists and draughtsmen.

First-hand accounts, diaries, journals, letters, logbooks, manuscripts, and publications by Darwin, FitzRoy and the senior officers and crew are used whenever possible to enliven and authenticate the series of stories. Almost all the engravings selected by FitzRoy for publication in the illustrated volumes of the *Narrative* (I and II) are featured in this publication. They were specially photographed and appear courtesy of the Royal Geographical Society.

For everyone associated with the voyage it provided life-changing experiences, yielding tales of endurance, remarkable seamanship, survey work, artworks and illustrations, as well as discoveries relating to the major branches of natural history. It ranks arguably as the most significant of all scientific expeditions.

PRINCIPAL PORTS OF CALL AND STOPOVERS

Devonport, Plymouth	set sail on 27 December 1831
Cape Verde Islands	16 January 1832
Bahia (San Salvador)	29 February 1832 & 1–6 August 1836
Rio de Janeiro	5 April–25 June 1832
Montevideo	26 July 1832
On the east coast of South America	February 1832–May 1834
Falkland Islands	March 1833 & March 1834
Strait of Magellan	January 1834 & May–June 1834
On the west coast of South America	June 1834–September 1835
Valparaiso	23 July–10 November 1834 & 11 March–27 April 1835
Copiapo	22 June–4 July 1835
Iquique	12–15 July 1835
Callao	19 July–6 September 1835
Galapagos Islands	15 September–20 October 1835
Tahiti	15–26 November 1835
Bay of Islands, New Zealand	21–30 December 1835
Sydney, Australia	12–30 January 1836
Hobart, Tasmania	15–17 February 1836
King George Sound, Western Australia	6–14 March 1836
Cocos Keeling Islands	1–12 April 1836
Mauritius	24 April–9 May 1836
Cape of Good Hope	31 May–15 June 1836
St Helena	8–14 July 1836
Ascension Island	19–23 July 1836
Bahia	1–6 August 1836
Azores	20–24 September 1836
Falmouth, Cornwall	2 October 1836

(Source: Dr Janet Browne, *Charles Darwin: Voyaging*, 1995)

Darwin's observations during the voyage lead to the bestselling publications *On the Origin of Species* (1859) and *The Descent of Man* (1871), which laid the foundation for modern evolutionary science. His revolutionary ideas shocked Victorian society as he dared to suggest that man was descended from apes, challenging the orthodox version of the Creation propagated through The Bible. The aim of this book is not to enter into the debate on evolution, which continues to this day, but rather to highlight the chain of events, the different characters and the climate of the times, that not only set *Beagle's* second voyage in motion, but made it so successful and of enduring fascination.

After FitzRoy's voyage HMS *Beagle* would undertake a third and final survey expedition in Australian waters. But what do we know about the ship that has shaped all of our lives?

Chapter 1
The Origin & Design of HMS *Beagle*

The reader will be surprised to learn that she (HMS Beagle) belongs to that much-abused class, the '10-gun brigs,' – coffins, as they are not unfrequently designated in the service, not-withstanding which, she has proved herself, under every possible variety of trial, in all kinds of weather, an excellent sea boat. – John Lort Stokes, *Discoveries in Australia; with an Account of the Coasts and Rivers explored during the Voyage of HMS Beagle in the years 1837-43* (1846).

On 16 February 1817 working drawings were sent to the Royal Naval Dockyard of Woolwich on the River Thames for the ships HMS *Beagle* and HMS *Barracouta* (one of *Beagle*'s sister-ships). *Beagle*'s keel was laid in June the following year and almost two years later she was launched on 11 May 1820 as a 10-gun brig, at a cost of £7,803. Some of her sister-ships were also built there; some at the Royal Naval Dockyards of Chatham, Deptford, Plymouth, Portsmouth and Sheerness; and others in private yards.

A period painting by Nicholas Pocock (1741–1821), a Bristol-born maritime artist, gives a vivid impression of the scale and enterprise of Woolwich dockyard towards the end of the eighteenth century. The dockyard was established in about 1514, founded by Henry VIII for the construction of his new ship the *Henri Grace à Dieu*, popularly known as 'Great Harry'. The yard was enlarged several times until the middle of the nineteenth century and Royal Navy vessels were constructed there until its closure in 1869.

Beagle was designed as a two-masted ship by Sir Henry Peake, the Surveyor of the Navy from 1806 to 1822, and was not originally intended as a survey vessel. Her rig was changed before the first survey voyage, and although Darwin described her as a brig, implying that she had two masts, she actually sailed as a barque, with three masts, on all her survey voyages.

Conclusive proof of her rig can be found in several of the officers' accounts who served on board. In *Life and Letters of the Late Admiral Sir B. J. Sulivan, KCB, 1810–1890*, written by his son, Henry Norton Sulivan, and published after the Vice-Admiral's death in 1896, Sulivan records that he was about to join the survey ship as a midshipman: 'December 11, 1827, was the day on which I first saw the *Beagle*, in which I served so long afterwards.

I find that date in my log: exchanged numbers with H. M. Barque *Beagle*: 0.40 A.M., anchored near H.M.B. Beagle'.

Robert FitzRoy recalled the ship's rig in the *Narrative*: 'She was rigged as a bark [barque]; her masts were strongly supported by squarer cross-trees and tops, and by larger rigging than usual in vessels of her tonnage'.

It was in 1807, during the Napoleonic Wars (1803–1815), that Peake created a smaller and cheaper type of naval ship of around 235 tons, which could operate in shallow and deep waters and carry carronades. A carronade was a short cast-iron cannon, developed by and named after the Carron Company, an ironworks in Falkirk, Scotland. This type of gun, nicknamed the 'smasher' or 'devil gun', caused extensive damage at short range, and was widely used by the Royal Navy until the end of the 1850s.

The carronade was a tremendous technological and economic advance. They were considerably lighter than conventional guns, and their effectiveness meant that fewer were required on smaller (and therefore cheaper) ships to produce comparable firepower to a much larger vessel. Even so, this effectiveness was limited to short range.

Peake's vessels, which included HMS *Beagle*, were known as the Cherokee, Cadmus and Rolla class. They are grouped together because of their similarities. The Royal Navy ships *Cherokee, Cadmus* and *Rolla* were all launched in 1807. *Cherokee* was built in the Blackwall yard and sold twenty years later; *Cadmus* was constructed at Deptford and was sold to the Coastguard in 1835; *Rolla* was built at Northfleet, near Gravesend, and was sold in 1822.

Royal Dockyard of Woolwich (1790), oil painting by Nicholas Pocock. As a master of a merchant ship he used to illustrate his logbooks with maritime scenes. He received encouragement from Sir Joshua Reynolds, and became one of the most prominent visual recorders of the French Revolutionary and Napoleonic Wars.

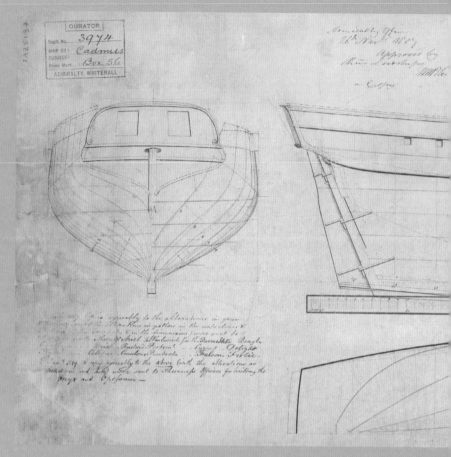

Plan of HMS *Cadmus*. The plan numbered 3974 was a copy, dated 26 November 1809, of the original 1807 plans for the 10-gun brigs. The inscribed notes provided additional information for Professor Keith S. Thomson to recreate the lines of the *Beagle* more accurately.

Beagle was one of more than a hundred vessels in this composite class. These ships were designed to carry eight 16-pounder or 18-pounder carronades and two long 6-pounder chase guns, although throughout Darwin's global voyage the vessel carried fewer guns, and only one 6-pounder carronade, which was prominently placed on a turntable at the forecastle. FitzRoy recalled the exact type and position of each gun: 'On the forecastle was a six-pound boat-carronade: before the chestree were two brass six-pound guns…. Abaft the main-mast were four brass guns, two nine-pound, and two six-pound'.

No original specific draft or plan exists for *Beagle*. Extensive research work by John Chancellor, Lois Darling, David Stanbury, Professor Keith S. Thomson, and ship modeller Karl Heinz Marquardt has greatly contributed to our understanding of the lines of the vessel and her appearance, above and below decks.

In Thomson's publication *HMS Beagle: The Ship That Changed The Course Of History* he reveals his initial frustrations in trying to gain information on drafts relating to the *Beagle* from the National Maritime Museum at Greenwich, the largest maritime museum in the world. He received the following reply: 'Unfortunately we have no model or plans of the…vessel, nor can we suggest any literature to help with your query'.

Fortunately his second enquiry proved to be more fruitful. The National Maritime Museum did have plans of sister-ships, such as *Cadmus*, prompting

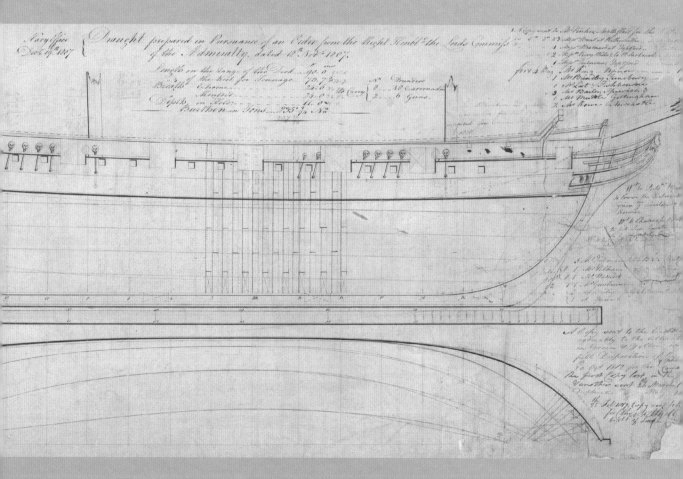

Thomson to arrange an appointment to view them. At least the sister-ships would provide the basic hull dimensions of the *Beagle*. Lois Darling had preceded Thomson on this quest to Greenwich, but Thomson was sceptical about the accuracy of Darling's reconstructions of the vessel's appearance.

After detailed examination of *Beagle*'s sister-ships over several days he viewed document numbered 3974 (current Museum reference ZAZ5137). It was a copy, dated 26 November 1809, of the original 1807 plans for the 10-gun brigs. It bore the Admiralty stamp *Cadmus* but other than that it was not clear that the plan related to any specific ship. Thomson observed that the hull lines had been altered several times in different coloured inks.

Thomson spotted, among the many annotations and notes dating from 1817 inscribed on the front of the sheet, the following words: '16 July 1817. Copies agreeably to the alterations in green and yellow (except the alterations in yellow in the water lines and stations of the forebody & in the dimensions) were sent to Deptford for the Alacrity & Ariel, to Woolwich for the Barracouta, Beagle'. He also observed an inscription on the reverse of the draft of what he believed was a complete list of other vessels that were subject to these design modifications. This evidence showed that the *Beagle* and several of her sisters had been altered from the original design for a 10-gun brig by having the bulwarks (sides of the ship above the upper deck) raised about 6 inches at the stem reducing to 4 inches at the stern. This was a significant breakthrough, enabling an accurate reconstruction of *Beagle*'s

HMS *Beagle*, port bow view of the ³⁄₁₆in:1ft scale model by Karl Heinz Marquardt, showing her fully rigged.

original hull (though not later modifications), and a strong starting point for the reconstruction of the ship.

The general arrangement of *Beagle*, the hull structure, fittings, armament, standing and running rigging, sails and boats, etc., have all now been recreated in convincing detail, albeit with some speculation. Karl Heinz Marquardt's publication *HMS Beagle: Survey Ship Extraordinary* (1997), reveals his dedication to create, using all the available information, the most realistic ³⁄₁₆in: 1ft scale model of the survey ship. It is now on display in the Deutsches Schiffahrtsmuseum (German Maritime Museum) at Bremerhaven, Germany.

Marquardt studied all the ship plans and contemporary artworks. In the 'Design' section of his book he noted that 'one has to consult the various artistic impressions by Augustus Earle and Conrad Martens, John Lort Stokes, Owen Stanley and Henry I Campbell, as well as written "Narrative's" and other general contemporary evidence'.

Additional information in the form of eye-witness accounts, artist watercolours, and drawings produced by the officers and seamen who served on board *Beagle*, as well as from sailors of other ships who came into contact with her, are all useful in building up a picture of the *Beagle*, but should be considered with caution.

Augustus Earle and Conrad Martens, the official artists on the second voyage, were not professional ship portrait painters or trained marine artists. Earle had a gift for drawing people, while Martens excelled at landscape painting. So historians have to be careful to balance what might at first glance appear to be a faithful reproduction of the lines and appearance of the survey vessel (or for that matter any maritime vessel they drew or painted) with the reality that the artists incorporated varying degrees of artistic licence into their work. From time to time FitzRoy advised his artists to alter details to achieve nautical accuracy in their preparatory pictures. For those illustrations he approved, he would initial them on the front. Some of their topographical images were sent to the Hydrographic Office for inclusion on the sheets of the Admiralty charts.

The painting 'HMS Beagle off Fort Macquarie, Sydney Harbour' is the best-known watercolour image of the survey ship, although it relates to her third and final expedition in 1837–43. It was created by Owen Stanley (1811–50), captain of HMS *Britomart*, a sister-ship of the *Beagle,* who was also an artist of promising ability and occasionally produced accomplished work. Stanley's picture featuring the *Beagle* was completed when *Britomart* came into contract with *Beagle* in Australian waters. It is loosely constructed and to a small scale (the size of a large postcard). It is full of atmosphere and charm and is now in the collections of the National Maritime Museum, Greenwich. However, his painting technique makes it difficult to determine if this really is a technically accurate portrayal of the *Beagle*.

A series of drawings created by Philip Gidley King, who was a midshipman on the second of the *Beagle* survey expeditions, was originally believed to be

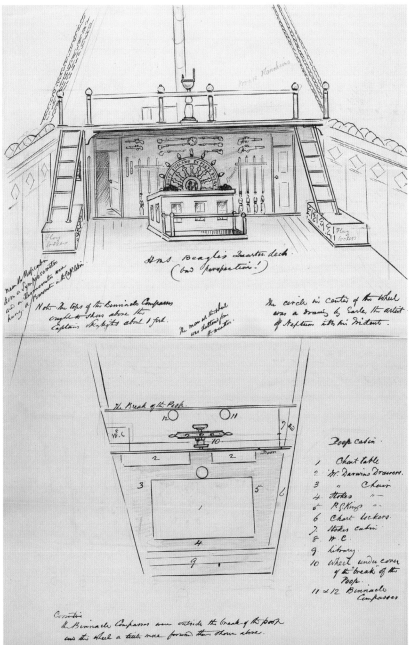

Above: 'HMS Beagle off Fort Macquarie, Sydney Harbour', watercolour by Captain Owen Stanley. Stanley entered the Royal Naval College, Portsmouth, aged 15, and in the 1820s served in the *Druid* and in the *Ganges*. He spent the next four years on the coast of South America, and in 1830 was employed under Captain Phillip Parker King in the *Adventure* (with the *Beagle*) on a survey of the Strait of Magellan. Stanley was elected a fellow of both the Royal Society and the Royal Astronomical Society.

Left: View towards the poop and interior of the poop cabin, memory drawings by Philip Gidley King. The top illustration shows the quarterdeck of the *Beagle* with Nelson's celebrated signal around the rim of the ship's wheel. The entrance to the added poop cabin was through the open door to the left of the wheel. John Lort Stokes slept in the small space to the immediate left of this door underneath the steps leading to the poop deck. The illustration below it shows the layout of the poop cabin. No. 1 indicates the position of the charting table; nos 3, 4 and 5 the placements of the chairs for Darwin, Stokes and King.

The Origin & Design of HMS *Beagle* 27

the most convincing, as they include a great deal of detail. However, they are not contemporary: they were reconstructed from memory by King more than fifty years after the second voyage for the publisher John Murray's first illustrated edition of Charles Darwin's *Journal of Researches* (better known today in the un-illustrated format as *The Voyage of the Beagle*), published in 1890.

The artworks for the illustrated edition were produced by the British artist, author and acclaimed gun-maker, Robert Taylor Pritchett (1828–1907). Despite royal support from Queen Victoria, who commissioned watercolours of ceremonies relating to her Golden and Diamond Jubilees, and impressive sailing experience (he completed two circumnavigations in private yachts: in Joseph Lambert's *Wanderer* and Sir Thomas and Lady Brassey's *Sunbeam*), his artworks lack the authenticity of FitzRoy's artists. They are more closely allied in spirit to the early illustrations of *The Boy's Own Paper*, the first issue of which was published in 1879. It regularly featured stories relating to seafaring, exploration and natural history.

Many of Pritchett's illustrations were inspired and adapted from FitzRoy's official artists, as well as from pictorial contributions and reminiscences of the officers and crew. He engraved two of Philip Gidley King's memory drawings: one portrays HMS *Beagle*'s 'Middle Section Fore and Aft' and the other the 'Upper Deck'. Both are dated 1832, although they were recreated many years later.

King also sketched another image (see page 27), a view towards the poop with an interior layout of the poop cabin, which did not feature in the illustrated publication. The original drawing is now part of the John Murray Archive. It provides some fascinating details, although it is unlikely that the firearms arranged behind the wheel would have been openly exposed to the elements.

Augustus Earle is known to have painted a picture of the Roman god King Neptune, ruler of the sea, with his trident on the hub of the ship's wheel during the early part of *Beagle*'s second voyage. Earle may well have been responsible for drawing the words 'England Expects Every Man To Do His Duty', a reduced version of Vice-Admiral Horatio Nelson's inspirational signal hoisted before the Battle of Trafalgar on 21 October 1805, around the wheel's rim.

Earle also created a lively and light-hearted image of the crossing-the-line ceremony, which took place when ships crossed the equator. The original drawing, or watercolour, has not been located but it was engraved as an illustration in the *Narrative*. It has become one of the best-known images of the voyage. The picture provides useful evidence of the appearance of the mid-ship section of the main deck. Earle portrays a skylight that is likely to be that of the gunroom.

Philip Gidley King created a fascinating drawing of the *Beagle*'s poop cabin that was used as the chartroom and berth of Charles Darwin (see page 30). Located on the quarter deck at the stern of the ship the cabin contained a large drafting table and provided accommodation for the surveying officers.

The dimensions have been estimated at 11 feet by 10 feet. King also drew a diagram of the lower deck indicating the arrangement of the officers' cabins. This is now part of a private British collection.

The illustration of the ship's poop cabin was probably drawn for Darwin and has pencil inscriptions in his own hand. It shows Darwin's chest of drawers, washstand, instrument cabinets, and book cases, oven, as well as the mizzen mast of the ship. The steering rope and tiller are clearly visible below the vast horizontal drafting table. The drawing is now in the Cambridge University Library. Darwin did not have exclusive use of this space. King also slept within the cabin and the drawing indicates that Stokes slept under the stairs beside the cabin.

On 25 April 1882 the *Morning Post* published Stokes' tribute to his cabin companion who had died that month:

> *…perhaps no-one can better testify to his [Darwin's] early and most trying labours than myself. We worked together for several years at the same table in the poop cabin of the Beagle…he with his microscope and myself with the charts. It was often a very lively end of the little craft and distressingly so to my old friend who suffered greatly from sea-sickness. After perhaps an hour's work he would say to me, 'Old fellow, I must take the horizontal for it'…and stretch out on one side of the table. This would enable him to resume his labours for a while when he had again to be down.*

'Crossing the Line' (17 February 1832), engraving after Augustus Earle from the *Narrative* (1839). Preparations for the ceremonies of crossing the line started on 16 February 1832. In Darwin's Diary he recorded on the 17th: 'We have crossed the Equator. I have undergone the disagreeable operation of being shaved'.

The Cambridge University Library contains a letter from Admiral Sir Bartholomew James Sulivan that includes a crudely drawn sketch of *Beagle*'s poop cabin/chartroom showing where Darwin hung his hammock. King's drawing in the John Murray Archive precisely indicates the position of his chair, as well as those of Darwin and Stokes (*Beagle*'s assistant surveyor) around the drafting table. King's drawing from the Cambridge University Library bears an inscription in Darwin's hand showing the location of Stokes's cabin under the stairs immediately outside the entrance to the poop cabin. From this evidence and the fact that King had an influential naval father (the commander of *Beagle*'s first expedition), it has been presumed that he enjoyed the priviledge of sharing the large cabin with Darwin (although at the time he was only a midshipman). Darwin himself makes no direct reference to King being his cabin companion.

If King did share the poop cabin with Darwin it is likely he would have slung his hammock on the starboard side of the cabin, while it is certainly known that Darwin positioned his to the port side close to his personal storage space and chest of drawers. Darwin was 6 feet in height and fortunately was a slim build. Every night in order to rig up his hammock he had to take out the top drawer from the chest of drawers in the forward bulkhead. Only two feet separated Darwin from the large skylight above. 'It is rather amusing,' he wrote in his Diary, 'whilst lying in my hammock to watch the moon or stars performing their small revolutions'.

Darwin had no prior seafaring experience (although he had made a Channel crossing for a visit to Paris) and revealed his ignorance in his Diary on 23 November 1831: 'My notions of the inside of a ship were about as indefinite as those of some men on the inside of a man, viz. a large cavity containing air, water & food mingled in hopeless confusion'.

In September 1831 Darwin noted in his Diary on first seeing the inside of the *Beagle* at Devonport: 'The absolute want of room is an evil that nothing can surmount'. He would later revise his view, deciding that the compact space was a help rather than a hindrance to his work, because everything was close to hand. Although Darwin was a prolific writer he left behind surprisingly little material relating to shipboard life, and no significant descriptions of cabins and living quarters onboard. One brief but evocative recollection noted in his Diary on 28 June 1832 conveys the sounds of shipboard life: 'It is something quite cheering to me to hear the old noises.– the men foreward singing: the centinel [marine guard] pacing above my head & the little creeking of the furniture in the Cabin'.

As the senior officer and to enforce discipline the captain lived apart from the other officers and crew. A strict shipboard hierarchy ensured that the senior officers'

Opposite: Poop interior showing Darwin's hammock, sketch within a letter from Admiral Sir Bartholomew James Sulivan to Sir Joseph Dalton Hooker written circa 1882. This pen-and-ink sketch indicates that Darwin slung his hammock on the port side of the poop cabin. The mizzen mast and part of the drafting table are also shown.

Below: Interior of the poop cabin, memory drawing by Philip Gidley King. Immediately in front of the 'Scale of Feet' can be seen four divided compartments. The larger section contained bookcases. Over the ship's steering gear was placed the rectangular drafting table. In front of the table the circular section indicates the position of the ship's mizzen mast. To the right, the open door of the cabin is visible, and to the immediate left the largest divided section visible (on the port side of the ship) was Darwin's chest of drawers.

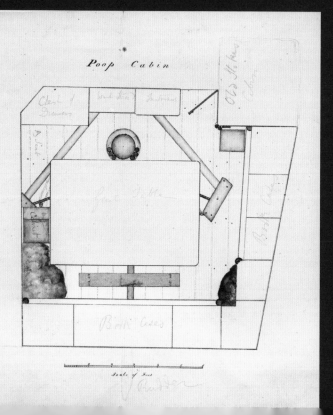

Hammock.

Books

Seat

drawers

Table

not said that the narrow
space at End of the chart table
was his only accomodation
for working, dressing, and
sleeping. the Hammock being
left hanging over his head
by day when the sea was
—at all rough. that he might
lay in it with a book in
his hand when he could
not any longer sit at the
table. his only stowage
for clothes being several
small drawers in the corner

cabins were positioned closest to the captain. Not surprisingly the captain's cabin was of a significant size and was the largest after Darwin's poop cabin.

In FitzRoy's *Narrative* he provides only a brief description of the ship's working and sleeping space. 'The skylights were large; there was no capstan: over the wheel the poop-deck projected, and under it were cabins, extremely small, certainly, though filled in inverse proportion to their size. Below the upper deck her accommodations were similar to, though rather better than those of vessels her class'.

One record that has survived was written by Conrad Martens and appears in his shipboard journal and also in his *Notes on Painting* (March 1834). He provides a rare insight into the appearance of one of *Beagle*'s cabins. It is likely that he had been allocated the berth of the assistant surgeon, as this position and related berth were vacant from April 1832 following Benjamin Bynoe's promotion to acting senior surgeon.

Martens observed:

> *I am now writing, in my cabin, which by the by I must tell you is allowed to be a pattern of neatness and convenience, the door which opens into the gunroom. It is lighted by two bulls eyes [glass prisms set into the deck to provide light] from the upper deck in the manner of a skylight, and as I am upon too familiar a footing with my messmates ever to think of shutting the door, a good deal of light comes in that way also. A tasty blue cloth curtain, however, is drawn at night, closing likewise a small window alongside of it. Facing the door, built in as it were occupying the whole length of the cabin, is a nest of drawers of 3 tiers, above which is the bed place, particularly well adapted for those who like to lie high, being at least 4½ feet from the deck.*

The artist also described the cabin's dimensions and the arrangement within:

> *The dimensions of the cabin is 6 feet long by rather more than 5 feet wide and 6 feet high. The bedplace is not very wide, being of course only intended for one person...On the left of the door is my table, desk lamp, and drawing materials. The end...is occupied by books, guns, pistols, my plate, a picture, and sundry other useful articles, arranged and fixed in such a manner that the outmost motion of the vessel will not disturb. The whole is painted in imitation oak...with the exception of the drawers, which are of mahogany...*

HMS *Beagle* had a length over her deck of 90 feet and her extreme breadth was 24 feet 6 inches. She was almost 10 feet shorter than Captain James Cook's converted collier brig HM Bark *Endeavour*, the discovery vessel which successfully completed a global circumnavigation in 1768–71. The *Beagle* was also smaller than Captain Matthew Flinders' sloop HMS *Investigator*, which charted many areas of the Australian coastline in 1801–3. *Endeavour* had a total of ninety-four men onboard; HMS *Investigator* a crew of seventy-eight; while under FitzRoy's command, *Beagle* had a complement of seventy-four.

It is remarkable that *Beagle* was selected for any survey voyage. She was of a class of ships that had gained a notorious reputation for poor manoeuvrability and for sinking. Around a quarter of the ships of her class were either lost at sea or wrecked, earning them the nicknames 'Coffin Brig' or 'Half Tide Rock'.

There were significant design faults with the *Cherokee*, *Cadmus* and *Rolla* class. The renowned British naval historian William James (1780–1827) was dismissive of them. He remarked in his *Naval History 1793–1827*, '[it is] surprising indeed that the navy-board should continue adding new individuals by dozens at a time...to this worthless class'.

The class had a low freeboard and solid bulwarks so the decks were easily awash and the water would not readily run off. Darwin wrote in his shipboard Diary on 20 April 1833: 'It blew half a gale of wind; but it was fair & we scudded before it.– Our decks fully deserved their nickname of a "half tide rock"; so constantly did the water flow over them'. However, many of the faults were, to a greater extent, addressed and rectified by FitzRoy. Her re-rigging as a barque for her first survey expedition made her easier to handle. For the second expedition the Admiralty paid for the major structural alterations, repairs and renovations, but FitzRoy also expended a significant amount of his own money to ensure that he had the best-provisioned survey ship.

He wrote in his *Narrative*:

> The Beagle was commissioned on the 4th of July 1831, and was immediately taken into dock [Plymouth] to be thoroughly examined, and prepared for a long period of foreign service. As she required a new deck, and a good deal of repair about the upper works, I obtained permission to have the upper-deck raised considerably [eight inches abaft and twelve forward], which afterwards proved to be of the greatest advantage to her as a sea boat, besides adding so materially to the comfort of all on board.

FitzRoy wrote that he was 'resolved to spare neither expense nor trouble in making our little Expedition as complete with respect to material and preparation, as my means and exertion would allow, when supported by the considerate and satisfactory arrangements of the Admiralty'.

Starboard cutaway view of HMS *Beagle* showing the interior arrangement of cabins and spaces, memory drawing, by Philip Gidley King. This illustration offers teasing glimpses into the main shipboard spaces, including the men's mess tables, the midshipmen's berth, the gun-room, captain's cabin (1), and added poop cabin (2), where the charting took place and Darwin slept.

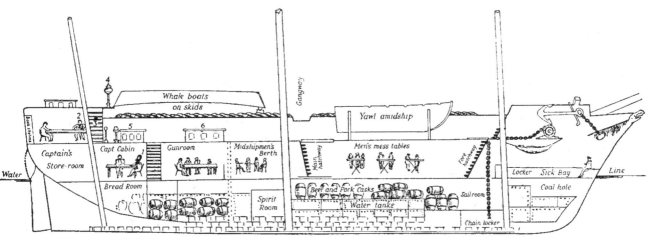

The cost of the alterations and improvements to *Beagle* was only £220 less than her original construction costs. Her timbers were rotten from her first voyage. She had to be largely re-built, strengthened, and her bottom re-coppered to stop the growth of seaweed and the destructive effects of the ship's worm, a mollusc of the tropics that readily bores into the hulls of wooden ships causing extensive damage.

These alterations were noted by FitzRoy:

> *While in dock, a sheathing of two-inch fir plank was nailed on the vessel's bottom, over which was a coating of felt, and then new copper. This sheathing added about fifteen tons to her displacement, and nearly seven to her actual measurement. Therefore, instead of 235 tons, she might be considered about 242 tons burthen...six superior boats (two of them private property) [besides a dinghy carried astern] were built expressly for us, and so contrived and stowed that they could all be carried in any weather.*

Beagle carried eleven anchors, far more than was usual for a 10-gun brig, but essential for effective survey work. Similarly Captain FitzRoy insisted on extra boats (seven in total), including four whaleboats, a cutter, a yawl and a dinghy (suspended astern), necessary for landing ashore, surveying and taking soundings in shallow waters and smaller tributaries, as well as exploring rivers. The yawl and cutter had respective lengths of 26 feet and 23 feet, while two of the whaleboats had lengths of 25 feet, and the larger two of 28 feet, probably paid for by FitzRoy. He knew that the ship's boats would speed up the onerous and time-consuming task of surveying. They were not a luxury; they were essential to ensure the success of the expedition.

Beagle also benefited from several new inventions. She had William Snow Harris's lightning conductors installed in her masts and spars; a new rudder fitted to the design of the naval captain, John Lihou (1792–1840); instead of the usual 'galley' fireplace, she had one of Frazer's stoves with an oven attached. On the first voyage the anchors had been taken in by Captain Phillips' patent capstan. FitzRoy had the capstan replaced with a new patent windlass. *Beagle* carried twenty-two chronometers that were carefully positioned in a dedicated space that was part of the captain's quarters. To ensure accuracy they were 'suspended in gimbals, as usual, within a wooden box', and these boxes were 'placed in sawdust, divided and retained by

Below left: HMS *Beagle*, ³⁄₁₆in: 1ft scale model by Karl Heinz Marquardt. In this photograph, the prominence of the ship's boats is evident in this deck view, which were vital for the survey work. Several were lost and destroyed on *Beagle*'s first expedition so FitzRoy wanted to be well prepared. In this photograph Marquardt has removed the foremast of the model to provide a clearer view of the foreward section of the ship. Note the carronade on the circular platform.

Below right: HMS *Beagle*, ³⁄₁₆in: 1ft scale model by Karl Heinz Marquardt. The stern of the ship is portrayed with survey boats in position, also the sounding winch, poop cabin skylight and, directly in front, the Azimuth compass. The poop deck was raised to accommodate the poop cabin, and this also made it an ideal platform to make observations.

partitions, upon one of two wide shelves'. FitzRoy's request to replace the iron cannons with brass ones to stop interference with his navigational instruments was turned down by the Admiralty: undeterred once the ship reached Rio de Janeiro he paid for replacements from his own pocket.

Interestingly Darwin, in a letter to his cousin, William Darwin Fox, in November 1831, states: 'Everybody, who can judge says it is one of the grandest voyages that has almost ever been sent out–. Everything is on a grand scale.– 24 Chronometers…'. Either FitzRoy had added a couple more precision timekeepers, or Darwin was mistaken. ·

George James Stebbing joined *Beagle* as FitzRoy's private assistant specifically to maintain the chronometers and other navigational instruments. The ship's roster records him as 'instrument maker'. He was the son of a mathematical instrument maker from Portsmouth who also sold instruments. Stebbing's father was awarded a royal appointment as Optician to the Duchess of Kent. FitzRoy insisted to the Admiralty that such a position was a necessity, although he personally paid Stebbing's salary of £300. The Admiralty agreed to bear the cost of feeding him. Entrance to the chronometer cabin was restricted to key personnel involved in their monitoring and maintenance.

John Lort Stokes acknowledged, on *Beagle*'s third (and last) survey expedition, that the survey ship 'proved herself under every possible variety of trial, in all kinds of weather, an excellent sea boat'. Perhaps this was due to her extensive design modifications. Certainly, Pringle Stokes had complained on the first voyage that 'Our decks were constantly flooded'.

In addition to the poop cabin Professor Thomson, and others, have noted the significance of a forecastle, which had also been fitted for the first voyage, that

> …*gave additional storage space for equipment and a place where the men could get out of the elements. Above all the forecastle deflected large seas shipped over the bow. They would break first on the forecastle deck and then be turned off to the sides. This made the whole main deck much dryer and safer. In comparison, in the open flush-decked brig of the original design every wave that came crashing over the bows was dumped straight onto the main deck.*

Credit must also be given to the professional skills of her officers and crew. HMS *Beagle* had some close escapes but she never foundered.

It has been suggested that the origins of *Beagle*'s name may derive from a dog-loving admiral, or an influential naval administrator who enjoyed hunting. However, it was not unusual for the Royal Navy to name ships after animals. Some of *Beagle*'s sister-ships were named *Badger, Ferret, Opossum, Reindeer, Sphinx, Weazle* and *Wild Boar*. However, the most popular category of names for those ships commissioned as survey vessels relates to birds. They included *Bustard, Curlew, Cygnet, Falcon, Kingfisher, Lapwing, Partridge, Pigeon, Skylark* and *Woodlark*.

There cannot be two Royal Navy ships in commission with the same name. So perhaps the name *Beagle* was the next one available on the list. She was the third vessel of the Royal Navy to bear that name. Coincidentally it was an apt name as the beagle dog has an excellent tracking instinct.

The first HMS *Beagle* was an 8-gun gallivat built around 1766 in Bombay. A gallivat was a small, armed vessel, powered by sails and oars, and found off the Malabar Coast in Southern India. The second vessel was a 18-gun brig-sloop of the Cruiser class (383 tons), constructed in 1808 at the Blackwall dockyard by Perry, Wells and Green. She served with distinction during the Napoleonic Wars, under Admiral Lord Thomas Cochrane (1775–1860) at the Battle of Basque Roads in 1809, and in other actions in Spain and France. She had been sold in July 1814, so the name became available once again.

Whether or not HMS *Beagle* had a dog as her figurehead is open to debate. Karl Heinz Marquardt offers two schools of thought: 'One...asserts that the utilitarian character of those small brigs, without real embellishment anywhere, warrants only a scroll whilst the other suggests a carved figurehead'.

The portraits of *Beagle* painted by the official artists are too sketchy to offer a conclusive answer. Captain Owen Stanley's watercolour 'HMS Beagle off Fort Macquarie, Sydney Harbour' suggests that her figurehead was of a dog shape, with an outstretched paw. Philip Gidley King's longitudinal sketch of the ship (the middle section fore-and-aft drawing), engraved by Pritchett, also suggests an animal, probably a beagle dog. However, no figurehead is present in Conrad Martens' engraved illustration 'Beagle laid ashore, River Santa Cruz', engraved as an illustration for the *Narrative* (see page 109) and this is the only image that presents a clear view of the ship's bow.

No portraits of the vessel prior to her survey voyages have been traced. She was, according to Darwin in a letter (written in stages) to his father from 8 to 26 February and 1 March 1832: 'not a particular ship'. In the early years there was no reason why she would attract any special attention. After she was launched she was placed 'in ordinary'; in other words she was waiting to be commissioned for duty. In July 1820 as part of the coronation celebrations of George IV (reigned 1820–30), HMS *Beagle* participated in a Royal Naval Review, and was allegedly the first naval vessel to sail fully under the old London Bridge.

On 7 September 1825 Commander Pringle Stokes was appointed captain. At last she had a role and was allocated to the surveying service under the management of the Hydrographic Office. As she was no longer required as an active fighting ship her armament was reduced to six guns. Twenty days after Stokes's appointment the survey ship was docked at Woolwich to effect repairs, and to be fitted out for her new role.

HMS *Beagle* completed her first four-year survey voyage returning to England on 14 October 1830. She was not the first choice for the second expedition. The Admiralty's preference was to commission one of her sister-ships, HMS *Chanticleer*. This vessel was built at Cowes and launched in 1808. She was refitted in 1827 as a survey ship and barque rigged. However, she was found to be in need of extensive repairs and, although HMS *Beagle* was not in prime condition, the Admiralty opted to commission her again. The *Chanticleer* was sold into the Customs service. She was renamed *W.V.5* (watch vessel) on 25 May 1863 and broken up in June 1871 at Sheerness.

Although HMS *Beagle* has overshadowed her sister-ships, several are worthy of mention. To date no detailed visual records of these ships have been traced. Very few were involved in any sea actions of significance. Some were converted into paddle-steamers, and many operated as packet ships conveying passengers and mail from Britain to the USA and Canada. But it is the contribution to exploration and surveying for the Hydrographic Office that defines these ships: *Barracouta, Britomart, Chanticleer, Fairy, Lyra, Saracen* and *Scorpion*, among others.

Construction of HMS *Barracouta* began at Woolwich dockyard in May 1820. She was commissioned as a survey vessel between 1821 and 1826. On 2 October 1821 she was commissioned under the overall command of Captain William Fitzwilliam Owen (1774–1857) in HMS *Leven* (a sixth rate of 20 guns) to participate in an extended expedition to survey parts of the eastern and western coasts of Africa, as well as the Arabian coast. Owen was actively involved in the fight against the African slave trade. In 1829 *Barracouta* was converted to a barque-rigged packet and sold in 1836.

The *Chanticleer* missed her chance of making history under FitzRoy and Darwin; however, she did excel as part of what became known as the 'Pendulum Expedition' in the 1820s under the leadership of Captain Henry Foster (1796–1831), a veteran polar explorer and gifted mathematician. This expedition was concerned with taking accurate observations of longitude and magnetic variations, winds and currents, the measurements of tides, and also to note the heights of coastal mountains and islands in the South Atlantic. All this data was collected to determine the true shape of the earth.

HMS *Fairy* was built at the Royal Dockyard of Chatham and launched on 25 April 1826. She worked as a survey ship from around 1832 to 1840, and eventually sank in the North Sea. Both *Saracen* and *Scorpion* were built at the Royal Naval Dockyard of Plymouth. The former was launched on 30 January 1831, and the latter in the following year. *Saracen* was active as a survey ship from 1854 to 1860, and then sold two years later, while *Scorpion* was commissioned for survey work between 1848 and 1858, and then sold to the Thames River Police and broken up in 1874.

HMS *Beagle* was designed as a small multipurpose ship of war, and later commissioned to undertake extensive survey work. Although conceived as an ordinary ship, through her association with Darwin and FitzRoy she is widely regarded as an extra-ordinary vessel, and one that has caught the public imagination. *Beagle* is now remembered as a ship that helped changed the course of history.

Chapter 2
Robert FitzRoy: 'Evolution's Captain'

He is an extra ordinary but noble character, unfortunately however affected with strong peculiarities of temper. Of this, no man is more aware than himself, as he shows by his attempts to conquer them. I often doubt what will be his end, under many circumstances I am sure it would be a brilliant one, under others I fear a very unhappy one.
– Letter from Charles Darwin to his sister Susan from Sydney, 28 January 1836

Robert FitzRoy by Samuel Lane. He is portrayed here in the uniform of a Vice-Admiral. However, Lane probably painted the portrait to mark FitzRoy's promotion to Rear-Admiral in 1857. As Lane died in 1859 it would seem that the portrait was updated by another artist when FitzRoy was finally promoted on the retired list to Vice-Admiral in 1863.

Vice-Admiral Robert FitzRoy (1805–1865) has been unfairly overshadowed by the achievements of Charles Darwin. Without FitzRoy's request to Captain Francis Beaufort, the Admiralty Hydrographer, for assistance to find him a naturalist, messmate and gentleman-companion, it is more than likely that Alfred Russel Wallace (1823–1913) would now be remembered as the man who conceived the notion of natural selection. Wallace developed his ideas independently of Darwin, although at a later date.

Darwin's shipboard position allowed him regular contact with FitzRoy. He was well placed to gauge the personality as well as the strengths and weaknesses of his captain. He thought: 'His greatest fault as a companion is his austere silence: produced from excessive thinking: his many good qualities are great and numerous: altogether he is the strongest marked character I ever fell in with'.

Darwin's letter to his father from the Cape Verde islands on 8 February 1832 is a good summation of FitzRoy's work ethic. 'I never in my life met with a man who could endure nearly so great a share of fatigue.– He works incessantly, and when apparently not employed, he is thinking. – If he does not kill himself, he will during this voyage, do a wonderful quantity of work'.

FitzRoy also exerted a great deal of effort in the production of the official *Narrative* of *Beagle*'s first and second voyages. The death of Pringle Stokes and the retirement of Admiral Phillip Parker King to Australia, who were her captain and expedition commander respectively during her inaugural survey voyage, meant that FitzRoy had to dedicate many months of his private time to work up their notes, logbooks and journals into a publishable format for the first volume. FitzRoy was entirely responsible for the second volume, and it reveals at times a highly accomplished literary facility.

Approaching Bahia in Brazil on 28 February 1832 FitzRoy provided a delightful description of their arrival:

As we sailed in rapidly from the monotonous sea, and passed close along the steep, but luxuriantly wooded north shore, we were much struck by the pleasing view. After the light-house was passed, those by whom the scene was unexpected were agreeably surprised by a mass of wood, clinging to a steep bank, which rose abruptly from the dark-blue sea, showing every tint of green, enlivened by bright sunshine, and contrasted by deep shadow: and the general charm was heightened by turreted churches and convents, whose white walls appeared above the waving palm trees; by numerous shipping at anchor or under sail; by the delicate airy sails of innumerable canoes; and by the city itself, rising like an amphitheatre from the water-side to the crest of the height.

Barbara Palmer (née Villiers), Duchess of Cleveland, by John Michael Wright (1617–1694). She was the favourite mistress of Charles II during the 1660s. She met the King shortly after her marriage to Roger Palmer in 1659. Robert FitzRoy was related to their son Henry FitzRoy, 1st Duke of Grafton.

Even so, FitzRoy's literary efforts were not rewarded by critical acclaim to the same extent as Darwin's. Darwin would have been delighted with the lengthy review relating to his contribution to the *Narrative* by the naturalist William Broderip in the *Edinburgh Review* (Journal 69), in which FitzRoy's volume was also praised for 'breathing a healthy spirit', and also the positive comments from a critic, probably also Broderip, that appeared in the *Quarterly Review* (Vol. LXV, December 1839 and March 1840) in which he was described as a 'first-rate landscape-painter with a pen' and that 'Even the dreariest solitudes are made to teem with interest'. The majority of Darwin's reviews were similarly positive, but the critics found many faults with FitzRoy's style and content. In private Darwin was also scathing about his captain's writing style. It is Darwin's third volume of the *Narrative* that is remembered today: FitzRoy's and King's volumes had a comparatively short print run.

FitzRoy's character was captured in a slight profile sketch, which was later turned into a print (see page 44). This revealing profile portrait is now attributed to the hand of Midshipman Philip Gidley King. It portrays a gentleman with an aristocratic air; a determined man of lofty principles, although with a highly strung demeanour. There is something troubling in this portrait. You can sense the emotional restraint. Darwin also described FitzRoy to his sister Susan as 'a slight figure, and a dark but handsome edition of Mr Kynaston, and according to my notions, pre-eminently good manners'. (Sir Edward Kynaston, Bart, 1775–1839 was known to the Darwin family as the vicar of Kinnerley in Shropshire.)

In Darwin's *Autobiography* he described FitzRoy's character as,

…a singular one, with many noble features: he was devoted to his duty, generous to a fault, bold, determined, indomitably energetic, and ardent friend to all under his sway. He would undertake any sort of trouble to

assist those whom he thought deserved assistance. He was a handsome man, strikingly like a gentleman, with highly courteous manners.

FitzRoy's forebears were indeed aristocratic. He was directly descended from Charles II's illegitimate son, Henry FitzRoy, the 1st Duke of Grafton, by the King's mistress Barbara Palmer (née Villiers), the Duchess of Cleveland. He counted dukes and lords on both sides of his family. But FitzRoy himself was untitled. He was born at Ampton Hall, Suffolk, in England, and from the age of 4 he lived at Wakefield Lodge in Northamptonshire, the grand Palladian-style mansion of the FitzRoy family.

Through his mother (his father's second wife), Lady Frances Anne Stewart (1777–1810), known as Fanny, FitzRoy was related to an aristocratic Anglo-Irish family. Fanny was the daughter of the 1st Marquis of Londonderry and half-sister of Viscount Castlereagh. As such, FitzRoy was also related to Robert Stewart, Lord Castlereagh (1769–1822), the British statesman and Chief Secretary for Ireland during the Irish Rebellion of 1798. His sensitivity to political and personal criticism led to his suicide. After being recommended rest by his doctors he slit his throat with a penknife in his dressing room.

Robert Stewart, 2nd Marquess of Londonderry (Lord Castlereagh), oil painting by Sir Thomas Lawrence (1769–1830). Known by his courtesy title 'Lord Castlereagh' he entered Parliament in 1795. He was captured on canvas (circa 1809/10) by Lawrence whose artistic skill and brilliant brushwork was widely recognized and attracted royal patronage. In 1820 he became President of the Royal Academy.

FitzRoy has left behind no written records regarding his concerns about his mental wellbeing before the voyage. But there are some intriguing words (albeit from a third party) in a letter from Professor John Stevens Henslow, Professor of Botany at the University of Cambridge (from 1827–61), to his favourite pupil Charles Darwin: '[FitzRoy] wants a man (I understand) more as a companion than a mere collector & would not take anyone however good a Naturalist who was not recommended to him likewise as a gentleman'.

These words have been put forward as part of the main argument to support the supposition that Darwin was selected by FitzRoy primarily as a form of 'mental insurance policy'. But they are ambiguous. It appears that FitzRoy was looking for someone of a comparable social class and of good manners. It is also possible that Henslow was trying to reassure Darwin (and more importantly his father) that he would be treated sympathetically on the voyage, and regarded as a social peer.

After FitzRoy's first meeting and interview with Darwin in September 1831 he wrote to a friend: 'I like what I see and hear of him, much…I will make him comfortable on board, more so perhaps than you or he would expect, and I will contrive to stow away his goods and chattels of all kinds'.

FitzRoy had originally written to Beaufort to seek assistance in finding him a naturalist for the voyage. Beaufort contacted George Peacock, Professor of

Mathematics at the University of Cambridge, who in turn wrote to Henslow, for help to fulfill FitzRoy's request. Peacock wrote in early August 1831:

My dear Henslow,

Captain Fitz Roy is going out to survey the southern coast of Terra del Fuego, & afterwards to visit many of the South Sea Islands & to return by the Indian Archipelago: the vessel is fitted out expressly for scientific purposes, combined with the survey: it will furnish therefore a rare opportunity for a naturalist & it would be a great misfortune that it should be lost:

An offer has been made to me to recommend a proper person to go out as a naturalist with this expedition:- he will be treated with every consideration; the Captain is a young man of very pleasing manners (a nephew of the Duke of Grafton), of great zeal in his profession & who is very highly spoken of; if Leonard Jenyns could go [Jenyns was Henslow's brother-in-law], what treasures he might bring home with him, as the ship would be placed at his disposal, whenever his enquiries made it necessary or desirable; in the absence of so accomplished a naturalist, is there any person whom you could strongly recommend: he must be such a person as would do credit to our recommendation

Do think on this subject: it would be a serious loss to the cause of natural science, if this fine opportunity was lost
The ship sails about the end of Septr.

During *Beagle*'s first survey voyage FitzRoy had taken over as captain after Pringle Stokes took his own life. So FitzRoy was forewarned of the rigours and demands of surveying in South American waters, and also the responsibility and isolation of command as the ship's captain.

Insights into the difficulties of undertaking survey work in South American waters are revealed in *The Journal of HMS Beagle in the Strait of Magellan by Pringle Stokes, Commander RN 1827*, illustrated by Lieutenant Robert H. Sholl, now in the Hydrograhic Office archive. On 1 February 1827 Stokes wrote:

…during the remainder of our cruize we had a constant heavy gale from the WNW, with thick weather & incessant, drenching, rain: the consequent discomfort in an open boat will be readily enough conceived. Throughout that interval – five days – we were all constantly wet to the skin; repeatedly in doubling the various head lands, we were obliged – after hours of efforts to pull the boat ahead of the violent squalls and cross turbulent sea that opposed us – to desist for a time and seek rest and shelter in any little cove that chanced to be at hand. The nature of the coast was not such as to invite us much to go on shore; for it is either high steep rocks; or a narrow beach composed of knobs of granite attrited to roundness and slipperiness by the action of the tides, and skirted, almost at high water mark by a scarcely penetrable jungle – Every where an utter solitude.

FitzRoy had agreed that Darwin could pay for his own passage, as he would be travelling in a private capacity. Darwin was not subject to the usual naval chain of command and discipline, and was at liberty to leave the ship whenever he pleased. Therefore, FitzRoy could confide in him without undermining his authority. His own officers risked being disciplined if they spoke frankly to their captain, or FitzRoy chose to disapprove of their views. Darwin would dine on a daily basis with FitzRoy in the captain's cabin. This was an unusual and clearly premeditated arrangement, and to an extent supports the argument that Darwin was selected in a dual capacity as naturalist and captain's companion.

Although FitzRoy's father, General Lord Charles FitzRoy (1764–1829), was an army officer and several of his ancestors were military men, there were also notable naval relatives, including his uncle Lord William FitzRoy (1782–1857), who was an Admiral. His father had a distinguished career having been appointed aide-de-camp to George III. He later became MP for Bury St Edmunds. He actively encouraged his son to join the navy.

FitzRoy was only 4 years old when his mother died on 9 February 1810. For five years FitzRoy was sent away to Rottingdean School (near the city of Brighton and Hove), and later attended Harrow school before being admitted in 1818, at the age of 12, to the Royal Naval College at Portsmouth. So it is hardly surprising that a boy who lost his mother at such a young age, and who then was sent to boarding school and afterwards straight into the navy should develop a strong degree of emotional restraint. It was a natural self-defence mechanism.

In adulthood his letters to his dear sister Fanny occasionally reveal glimpses into his character and state of mind. One lengthy letter he wrote to her after the death of their father, completed at the beginning of October 1830, is telling:

> *You, and you only know I valued his slightest word, for, like your-self it is not natural to me to be much, or suddenly affected, outwardly, by what occurs. I have been working hard, and have run many risks during the last two years, & through all, I have been influenced by the thought that it would give him satisfaction. His approbation I looked to as the true reward of any hard times I might pass, and I thought little of any other person's.*

Fanny (1803–1878) had married well. Her name changed from Frances FitzRoy to Frances Rice-Trevor, and later Lady Dynevor, after marrying

View of Portsmouth, circa 1815, lithograph by an unknown artist/printmaker. This lively, small-scale print shows a view of part of the Royal Dockyard of Portsmouth with sailors engaging in exercises and racing. HMS *Victory* can be seen to the far right. In 1812 the *Victory* had been placed in reserve in Portsmouth: she would not fight again.

George Rice-Trevor, 4th Baron Dynevor (1795–1869), who was Tory MP for Carmarthenshire from 1820 to 1831 and from 1832 to 1852. She tried her best to offer emotional support to her distinguished brother, and was a frequent and welcome visitor to the Fuegians during their residence in Essex. They called her 'Capen sisser' (Captain's sister).

The Royal Naval College in Portsmouth was formally established in 1729 and was originally called the Royal Navy Academy for the 'better education and training of up to forty young gentlemen [a year] for H.M. Service at sea'. The Academy opened in 1733, and by 1806 the number of places was increased to seventy and the name changed. By this time forty places were available for the sons of naval officers free of charge.

It is likely that FitzRoy's emotional restraint precluded him from writing at length and in detail about his experiences as a student. However, a charming letter written to his father from the College, on 20 May 1812, offers a glimpse of his pride in his achievements to date, and is indicative of youthful zeal and ambition: 'We had another examination last Monday & I took three more places which were all that I could take for I am now at the head of the part of the college…'. Towards the end of the letter he circled the following words: 'You must excuse it [the letter] for being hardly readable for I am in an enormous hurry'.

Fortunately life at the College has been described by Vice-Admiral Bartholomew James Sulivan (himself the son of a naval captain and grandson of a naval admiral) in his son's biography of him, *Life and Letters*. He was considerably less reticent than his fellow collegian, who was five years his senior. Although they did not train together at the College, both men were to serve together on the first and second of *Beagle*'s voyages.

Collegians who excelled at their shore studies and demonstrated flair and practical skills afloat, would almost certainly benefit from rapid promotion. The 'first medal' winner would be guaranteed automatic promotion from Midshipman to Lieutenant. Although the course was, in theory, three years, the students were actively encouraged to complete it as quickly as possible. The first to finish the course satisfactorily would receive the premier medal. FitzRoy was the star student of his intake being 'awarded full-numbers' (full marks in his exams).

In 1824, several years after FitzRoy, Sulivan was awarded the 'second medal'. Although James Inman, the College Professor, had wanted to award joint 'first medals', the Admiralty had refused his request. The 'first medal' was awarded to Inman's son, Richard. Even-handed Sulivan thought it was a fair result. He would later be awarded full marks for his Lieutenant's examination. In his autobiography he recalls the faculty of the College:

> The head of the studies was the Reverend Professor James Inman, D.D., author of the work on navigation, under whom were three assistant-masters for mathematics: first, Peter Mason, M.A.; second, Charles Blackburn, M.A., and third, Mr. Livesay. The preceptor, the Rev. W. Tate, M.A., took the classics classes, history, geography, and English. French was taught by M. Creuze, a French émigré.

Sulivan reveals that the cadets were also taught
...fencing and dancing. The forenoons were given to mathematics, the afternoons to French and drawing, the latter taught by a very superior master, Mr J. C. Schetky [John Christian Schetky, the marine painter]. There were also classes in naval architecture, which were taken by Mr. Fincham, the master-builder of the dockyard. We began geometry with Mr. Livesay but no boy could get on unless he studied in his own cabin and at the dining-room tables in the evenings.

Before FitzRoy took temporary and then full command of the *Beagle* he already had experience in South American waters, serving initially as a 'college volunteer' in the autumn of 1819, and a year later as a midshipman in the *Owen Glendower*. He was in good spirits writing home to Fanny: 'I am very happy & comfortable on board & am sure I shall like sea life very well'. He also indicated to her his interests and studies in various languages including Spanish, Italian and French.

Clearly FitzRoy had adapted well to the cramped and confined conditions of life afloat. He would serve onboard the *Owen Glendower* for more than two and half years. In this vessel he would receive real rather than theoretical naval training. He would be subjected to life-threatening risks, and ordered to climb aloft to set, furl and unfurl the ship's sails. If you could not lead by example you were doomed to fail in your command. (Coincidentally FitzRoy and *Beagle*'s first captain, Pringle Stokes, served together on this ship.)

FitzRoy was promoted to Lieutenant on 7 September 1824. He subsequently served onboard HMS *Hind* in the Mediterranean. FitzRoy and Sulivan would first serve together in the frigate HMS *Thetis* (fifth rate of 46 guns) in the Mediterranean and in South American waters under the enlightened and inspirational Captain Sir John Phillimore (1781–1840). Although the captain was initially sceptical of the College-educated cadets aboard his ship, he was soon swayed by their commitment, diligence and skills, and became an adamant supporter of them. Land-based Naval Colleges providing accelerated promotional prospects for exceptional Collegians but provided training for only a small number of cadets. Traditionally the majority of officers learned most of their skills afloat, working their way up through the ranks, so not surprisingly there was a degree of jealously and rivalry between the traditional and College-trained officers.

As a junior lieutenant onboard the *Thetis* Sulivan believed that,
...FitzRoy was one of the best officers in the service, as his subsequent career proved. He was one of the best practical seamen in the service, and possessed besides a fondness for every kind of observation

Above: Captain Robert FitzRoy, graphite heightened with wash by Philip Gidley King, dated 1838. Admiral Phillip Parker King also made a pencil sketch of FitzRoy, circa 1838. These portraits form part of an extensive album of drawings and prints (1802–1902) assembled by several members of the King family.

Opposite bottom: HMS *Ganges*, watercolour by Rear-Admiral Sam Hood Inglefield (1783–1848). *Ganges* is shown sailing out of Rio de Janeiro. Built of teak at Bombay Dockyard by Jamsetjee Bomanjee Wadia, she was launched on 10 November 1821. She was flagship of the South America Station under Admiral Otway for three years. In 1905, after a lengthy service career in the Mediterranean and Pacific she became part of the Royal Naval Training Establishment, and was finally broken up in Plymouth in 1930.

useful in navigating a ship. He was very kind to me, offered me the use of his cabin and of his books. He advised me what to read, and encouraged me to turn to advantage what I had learned at college by taking every kind of observation that was useful in navigation.

Writing to Fanny, FitzRoy described the interior of his cabin on *Thetis* and the alterations he had arranged. He had made his

> *…Cabins most comfortable – quite a little Paradise – I [have] been sadly expensive in the book line & have I flatter myself a complete library in miniature (excepting one or two works which I have ordered). You will hardly think that in a place 6½ feet square I stow – a broad Chest of Drawer 1 trunk – a large table – washing stand 1 or 2 hats 2 cloaks, sticks & umbrellas… & Guns, and upwards of four hundred volumes!*

Clearly FitzRoy was an avid reader.

FitzRoy was re-appointed as one of the lieutenants to the *Thetis* under a new captain, Arthur Batt Bingham (1784–1830). Sulivan had decided to stay with FitzRoy and on 11 December 1827 they encountered HMS *Beagle* on her first survey voyage. According to FitzRoy, Bingham was 'very kind *hearted*' but also one of the 'emptiest *headed* vain men that ever annoyed his subordinates with fidgety nonsense'. *Thetis*'s duties eventually led her into South American waters to join the ships under the command of Admiral Sir Robert Waller Otway (1770–1846). She was part of a strategic political process of securing a British presence in the area to maintain cordial relations, and develop economic and trading partnerships.

Below: Admiral Sir Robert Waller Otway, 1st Baronet, lithograph, by Maxim Gauci (c.1810–1846). Otway was born in Tipperary, Ireland. During a distinguished naval career, he saw action at The Glorious First of June, 1794, and with Nelson at the Battle of Copenhagen, 2 April 1801. Nelson described him as being as 'good as gold'. In 1826 Otway was knighted and sent as commander-in-chief of the South American station. He supported the Brazilians during the Brazilian War of Independence. After retiring from the navy he was appointed Groom of the Bedchamber to William IV and later Queen Victoria.

'Port Famine', engraving after Phillip Parker King from the *Narrative* (1839). Port Famine (now called Puerto Hambre) is located on the west side of the Strait of Magellan. Tragically, in August 1828, it is where Pringle Stokes committed suicide.

In August 1828 FitzRoy was appointed to HMS *Ganges* (second rate, 84 guns) as flag-lieutenant to Otway. FitzRoy's work ethic caught the attention of the admiral and he became a favourite, leading ultimately to his appointment to take over command of the *Beagle*.

At that time Pringle Stokes (1793–1828) was *Beagle*'s captain with the survey voyage under the overall command of Phillip Parker King (1791–1856) in HMS *Adventure*. There was nothing exceptional about Stokes's service record: perhaps he was regarded as a 'steady pair of hands' and therefore considered a good choice as captain for the voyage. But by mid-1828 Stokes could no longer cope with the demanding working conditions, the seemingly unending task in unforgiving and abysmal sea conditions. Some of his men were suffering from scurvy, he was overwhelmed by the survey challenges, and in the desolate waters of Tierra del Fuego he fell into a deep depression.

At Port Famine in the Strait of Magellan Stokes locked himself in his cabin for fourteen days, and then on 2 August shot himself in the head, although he did not die immediately. Laird Clowes' classic work *The Royal Navy: A History From the Earliest Times to the Present*, published in the last year of Queen Victoria's reign, provides only the briefest of details relating to Stokes's death. The language is telling: 'At length, worn out by toil and overwork, Stokes, of the *Beagle*, succumbed…'.

It would take ten days for Stokes to succumb. During that agonizing period it appeared, at least to Stokes, as if he might be recovering, but he died on 12 August. He was buried at Port Famine in a small burial mound that was used as the resting place for all who had died in that area. In Llewellyn Styles Dawson's *Memoirs of Hydrography*, published in 1885, he claimed Stokes was the 'victim to over exertion, worry and excitement attendant upon surveying labour in that part of the world'.

No obituaries appear to exist for *Beagle*'s first captain. The Admiralty were probably ashamed of his failure to command, his dereliction of duties and dramatic demise. However, on 8 January 1862, on the front page of *The Times* newspaper, the 'Military And Naval Intelligence' column reported that 'the Calypso, sailing corvette of 16 guns, Capt. Frederick B. Montresor, arrived at Spithead yesterday morning from the Pacific'. She had anchored in Port Famine on 19 October 1861 and there, 'the grave of Captain Pringle Stokes…was visited and put in order'. His grave is now marked by a simple Roman cross and inscribed: 'In Memory of Commander Pringle Stokes H.M.S. Beagle who died from the effects of the anxieties and hardships incurred while surveying the western shores of Tierra del Fuego 12.8.1828'.

The original cross can now be seen in the Salesian Museum at Punta Arenas in Chile.

FitzRoy was generous in his praise for his fellow officer. FitzRoy was a practising Christian and he wrote in the first volume of the *Narrative* some respectful words for his deceased colleague: 'Thus shockingly and prematurely perished an active, intelligent, and most energetic officer, in the prime of his life'. He continued: 'His remains were interred at our burial-ground, with the honours due to his rank, and a tablet was subsequently erected to his memory'.

Eight days before he died Stokes had drawn up his will on board the ship. It was signed by Phillip Parker King and witnessed by Lieutenant William George Skyring (c.1780–1833). There is a certain irony to the standard introductory statement that he was 'in sound mind and body'.

Stokes had suffered terribly with a raging delirium, although he had managed to recount his version of events, and provide a summary of *Beagle*'s voyage thus far to several officers during intermittent periods of lucidity. He had failed to keep the ship's journals and logs up to date. Successive officers and crew of *Beagle* would achieve considerably more in terms of exploration, discovery and surveying.

After Stokes's death, Skyring, as the ship's senior officer, took over temporary command of the *Beagle*. He successfully navigated the vessel from the Strait of Magellan back to Rio de Janeiro, arriving in October 1828. Here repairs were made to the ship and fresh provisions procured. He was more than worthy to succeed Stokes as the official commander of the ship, but instead Admiral Otway appointed FitzRoy.

Last Will & Testament of Pringle Stokes. He was almost certainly born in the town of Chertsey in Surrey, where he was baptised on 2 May 1793. Stokes joined the Royal Navy at the age of 12. He served in several ships, and gained experience in the Baltic, East Indies, and on the African Station. On 10 November 1822 he served with Captain Sir Robert Mends in the *Owen Glendower*. Before his appointment to the *Beagle* Stokes was promoted Commander in December 1823.

It is commendable that Skyring neither complained nor, it seems, harboured any grudge to this preferential appointment. However, he did receive a consolation prize and was placed in command of the second support vessel *Adelaide*. FitzRoy was well aware of the situation and later named, probably as a gesture of his gratitude for Skyring's good grace, one of two extensive South American sea lakes (around 40 miles long) Skyring Water. He named the other Otway Water, after his patron.

Writing to Fanny from Rio de Janeiro on 23 November 1828 FitzRoy was clearly delighted: 'It is not only promotion, but employment and that of the most desirable kind, for it opens a road to credit and character, and farther advancement in the Service. Providing I do not fail in my exertions'.

Phillip Parker King, by Thomas Woolner (1825–1892), bronze and black painted plaque, dated 1854. Woolner, sculptor and poet, was a founding member of the Pre-Raphaelite Brotherhood. He also designed a Wedgwood medallion of Charles Darwin in 1869, and a marble bust of him in the following year.

Unlike Stokes, FitzRoy revelled in the challenge and during his first command he could clearly sense a positive outcome through the wind and rain, and rough, treacherous waters of South America. Captain FitzRoy continued the survey work of the eastern coast and the series of channels comprising the Strait of Magellan where he would have his first encounter with Fuegians: 'Their figures reminded me of drawings of the Esquimaux, being rather below the middle size, wrapped in rough skins, with their hair hanging down on all sides, like old thatch, and their skins of a reddish brown colour, smeared over with oil, and very dirty'.

This part of the voyage would set off a chain reaction that would lead eventually to *Beagle*'s second survey expedition. On 28 January 1830 they anchored off the Brecknock Peninsula in the south west of Tierra del Fuego and on the following day FitzRoy ordered Matthew Murray, the *Beagle*'s master, with six men, to take one of the whaleboats to survey a headland that had earlier been named Cape Desolation by Captain Cook.

Early in the morning of 5 February FitzRoy received the news that Murray's whaleboat had been stolen. FitzRoy recounted the story in the *Narrative*:

Mr. Murray reached the place, and secured his party and the boat in a cave near the cape; during a very dark night, some fuegians, whose vicinity was not at all suspected, approached with the dexterous cunning peculiar to the savages and stole the boat. Thus deprived of the means of returning to the Beagle, and unable to make their situation known, Mr. Murray and his party formed a sort of canoe, or rather basket, with the branches of trees and part of their canvas tent, and in this machine three men made their way back to the Beagle, by his directions: yet, altogether favored by the only fine day that occurred during the three weeks that the Beagle passed in Townsend Harbour, this basket was twenty hours on its passage. Assistance was immediately given to the Master and the other men, and a chase for our lost boat begun, which lasted many days, but was unsuccessful in its object, although much of the lost boat's gear [including chronometer, theodolite, and other instruments] was found...

FitzRoy was incensed and during the protracted search-and-rescue operation three Fuegians were taken, and then a fourth, to act as guides, interpreters and hostages to effect the return of the stolen boat. The Fuegians were named in part after where and how they were taken. On 28 February York Minster was taken in Christmas Sound, a place FitzRoy had selected to build a replacement boat. Captain Cook had stopped there over the Christmas period on his second voyage of discovery (1772–5), hence the name. The outline of a nearby towering rock on Waterman Island had the appearance (in Cook's opinion) of York Minster and this inspired FitzRoy to name the man after that historic ecclesiastical building. On the following day a girl of around 8 years old was brought on board and was christened Fuegia Basket by the seamen, alluding to the makeshift craft that sailed from Cape Desolation. About a week later Boat Memory joined the other hostages. His name derived from his apparent knowledge of what had happened to *Beagle*'s boat. Later, on 11 May during another surveying trip, Jemmy Button was acquired and called after the 'large shining mother-of-pearl button' given to him to entice him to go onboard the ship.

Unsuccessful in recovering the whaleboat, FitzRoy decided to resume his original orders and changed course to rendezvous with his commander, Phillip Parker King, at Rio de Janeiro. In his *Narrative* FitzRoy reveals his plans for the Fuegians:

Fuegian of the Yapoo Tekeenica tribe, engraving after Conrad Martens from the *Narrative* (1839).

> *I had previously made up my mind to carry the Fuegians, whom we had with us, to England; trusting that the ultimate benefits arising from their acquaintance with our habits and language, would make up for the temporary separation from their own country. But this decision was not contemplated when I first took them on board; I then only thought of detaining them while we were on their coasts; yet afterwards finding that they were happy and in good health, I began to think of various advantages which might result to them and their countrymen, as well as to us, taking them to England, educating them there as far as might be practicable, and then bringing then back to Tierra del Fuego.*

FitzRoy also conveyed the body of a dead Fuegian shot by the crew of the *Beagle* at March Harbour, Christmas Sound, Tierra del Fuego, parts of two others, and 'the prepared skin of the head' of another. These were described by John Wilson, the *Beagle*'s surgeon in his autopsy report that features in a reduced version in the Appendix of the *Narrative*. Wilson's more expansive and revealing original manuscript (18 pages) is now part of the library collections of the Royal College of Surgeons (see William Clift's Diary, 30 November and 11 December 1830).

On 12 September FitzRoy wrote to King to inform him that he taken four Fuegians on board the *Beagle*. He listed their names and estimated ages as 'York Minster 26, Boat Memory 20, James [Jemmy] Button 14, and Fuegia

Basket (a girl) 9'. The real names of three of them were recorded in the *Narrative* as Yokcushlu (Fuegia), Orundellico (Jemmy), and El'leparu (York). Sketches of these three by FitzRoy appear in the second volume of the *Narrative*. Boat Memory was not included. All the Fuegians were inoculated against smallpox at the Royal Hospital in Plymouth upon their arrival in England and Boat Memory had an adverse reaction and died. FitzRoy felt personally responsible for his death. But he now had three Fuegians to accommodate and educate.

The Church Missionary Society came to his aid and provided accommodation and education at an infant school run by the Reverend William Wilson at Walthamstow in Essex. The education offered was rudimentary and included the 'plainer truths of Christianity', English lessons, 'the use of common tools, a slight acquaintance with husbandry, [and] gardening'.

FitzRoy was protective of the Fuegians and carefully vetted curious visitors. He did not want them to become 'curiosities' and it is to his credit that he did not exploit them for economic advantage. Captain Cook's second voyage had brought the first Tahitian to Britain. Mai, or Omai as he was called in England, benefited from the protection of Sir Joseph Banks. Mai became a celebrity and was in demand at fashionable dinner parties and social events. He was eventually returned to Huahine, in the Tahitian island group during Cook's third voyage (1776–80). But before Mai's departure he was captured on canvas in oils by, among others, Sir Joshua Reynolds (1723–1792), the first President of the Royal Academy of Arts, also by William Parry (1745–1791), one of Reynolds' favourite pupils, and William Hodges (1744–1797), the official artist of Cook's second voyage.

Strangely there are no extant oil portraits of the Fuegians. The only recorded images of them are known through the modest small-scale engravings after FitzRoy's drawings published in the *Narrative*.

In the late summer of 1831 FitzRoy and the Fuegians were summoned to St James's Palace for an audience with William IV and Queen Adelaide. The Queen offered one of her own bonnets and a ring to Fuegia Basket. She and the young teenager Jemmy Button had responded well to their English education, but York Minster who was older, probably in his early twenties, was far from a model pupil. It has been suggested that York Minster had formed an intimate relationship with the child Fuegia Basket, although the source for this derives from an imaginary account in *Cape Horn* (1939) by the seafarer Felix Reisenberg.

It is certainly true that FitzRoy had intended educating the Fuegians in England for two or three years and so something precipitated his hasty exit strategy. The Admiralty initially promised him a ship to return the Fuegians to Tierra del Fuego, but then reneged. FitzRoy engaged influential friends and family members to petition the Admiralty to honour their original promise, and they were successful. But it is doubtful it would have happened without the support of Captain Beaufort, the Admiralty Hydrographer, a kindred spirit who recognized the benefits of a follow-up voyage to complete previous survey work in and around the Fuegians' home waters. Beaufort was largely responsible for drawing up FitzRoy's survey instructions, although they went under the grand heading 'By the Commissioners for executing the office of Lord High Admiral of the United Kingdom of Great Britain and Ireland, &c.'.

During the agonizing wait for a decision from the Admiralty FitzRoy took matters into his own hands and contracted the merchant ship *John of London* to return his Fuegians. He ended up paying the larger part of £1,000 (the deposit) for a ship he never commanded. But FitzRoy was eventually appointed to command the *Beagle* again. He may have won the first round with the Admiralty through connections and influence, probably through the exertions of the 4th Duke of Grafton, and Lord Londonderry, Castlereagh's half-brother (who after the *Beagle* expedition secured for FitzRoy the position of MP for Durham). But the Admiralty were insistent that FitzRoy would receive no further funds above and beyond what was initially agreed to assist him on this voyage.

FitzRoy's expedition was the first to set off with no prospect of any officially sanctioned and paid for support ships or smaller survey vessels. *Beagle* had two support vessels on her first voyage, *Adventure* and *Adelaide*, a point unsuccessfully argued by Beaufort to support FitzRoy's initiative at hiring two vessels and purchasing a further two during the course of the second voyage. This does lend credence to the likelihood that the Admiralty had bowed reluctantly to FitzRoy's collective family influence for a ship, but beyond the repair and renovation of *Beagle*, her provisioning and fitting out, they were unwilling to commit to any further requests for financial support. The additional vessels that FitzRoy hired from his own pocket were two small 'decked-boats' of 15 and 9 tons, the *Paz* and *Liebre*. Darwin wrote in his Diary on Sunday, 14 October 1832: 'the Schooners came down from the creek & anchored alongside.– Their appearance is much improved by their refit; but they look very small'. Darwin's estimate of the vessels' size differs from others:

> *La Paz is the largest carrying 17 tuns; La Lievre [Liebre] only 11½– Between the two they have 15 souls.– Mr Stokes & Mellersh are in La Paz; Mr Wickham & King in the other.– They sail on Wednesday; I am afraid the whole party will undergo many privations; the cabin in the smaller one is at present only 2 & ½ feet high! Their immediate business will be to survey South of B. Blanca: & at the end of next month we meet them at Rio Negro, in the bay of Blas…*

FitzRoy also purchased the schooner *Unicorn* of 170 tons, which he renamed *Adventure*, and which he was later ordered by the Admiralty to sell. He would also benefit from the goodwill of Don Francisco Vascunan, who provided the short-term loan of the *Constitucion* of 35 tons, a vessel which again would be purchased from FitzRoy's own pocket at a cost of £400, to assist the surveying work before FitzRoy completed his chain of chronometric measurements.

HMS *Beagle* was officially commissioned on 4 July 1831. Her captain spared no expense and was adamant in his opinion that,

THE HOPE, IN THE STRAIT OF MAGALHAENS.

THE ADELAIDE, IN HUMMING BIRD COVE.

P. P. King.   S. Hall.

DISTANT VIEW OF MT SARMIENTO.

Never, I believe, did a vessel leave England better provided, or fitted for the service she was destined to perform, and for the health and comfort of her crew, than the Beagle. If we did want for any thing which could have been carried, it was our own fault; for all that was asked for, from the Dockyard, Victualling Department, Navy Board, or Admiralty, was granted.

FitzRoy still had to fight his corner to ensure that some of his demands were met at the outset. The Admiralty thought that five chronometers would be more than sufficient for the voyage but FitzRoy had other ideas and in fact took twenty-two precision timekeepers to ensure that longitude could be determined accurately. He paid £300 for these additional timekeepers and other instruments, and borrowed others. He wrote: 'Considering the limited disposable space in so very small a ship, we contrived to carry more instruments and books than one would readily suppose could be stowed away in dry and secure places; and in a part of my own cabin twenty-two chronometers were carefully placed'.

Beagle had been adapted and the height of her main deck enhanced and yet she was severely lacking in space. One of the main reasons related to the presence of the three Fuegians and the Reverend Richard Matthews, who would help settle them and establish a missionary station in Tierra del Fuego. The ship also had to carry all their gifts, provisions and equipment. Among the many useless items were dinner and tea services, presumably intended to assist the civilizing and gentrification process of families and friends.

Although the Fuegians were the major factor that precipitated the second *Beagle* voyage, there are very few descriptions of them onboard. All three had adopted fashionable English dress and had learned enough of the English language to converse, albeit at a rudimentary level. They adopted the manners, posture and deportment of Western Europeans. Jemmy Button liked to wear a cravat, gloves, and have his hair cut short. He regularly admired himself in the mirror. But these were only temporary transformations. Once settled they would gradually abandon their European clothes and table manners.

Darwin provides good descriptions of the three Fuegians:

[York Minster was a] full-grown, short, thick, powerful man: his disposition was reserved, taciturn, morose, and when excited violently passionate; his affections were very strong towards a few friends on board; his intellect good. Jemmy Button was a universal favourite, but likewise passionate; the expression of his face at once showed his nice disposition. He was merry and often laughed, and was remarkably sympathetic with anyone in pain: when the water was rough.

Jemmy Button was supportive of Darwin, who suffered terribly from seasickness throughout the voyage.

Jemmy was short, thick, and fat, but vain of his personal appearance; he used to wear gloves, his hair was neatly cut, and he was distressed if his well polished shoes were dirtied. He was fond of

Opposite: The *Adelaide* (centre) in Humming Bird Cove, engraving after Phillip Parker King from the *Narrative* (1839). *Adelaide* served as a support vessel on the *Beagle*'s first survey expedition.

This page: Detail of letter, 'My schooner is sold…' from Robert FitzRoy to Captain Francis Beaufort revealing FitzRoy's frustration at the British Admiralty's decision not to support his decision to acquire an additional vessel to speed up the survey work and reimburse him for the monies he had expended.

Opposite: Adventure at Port Desire, water-colour by Conrad Martens. The Beagle and Adventure arrived on 23 December 1833 at Port Desire (Puerto Deseado) on the coast of Patagonia. The port was named after Sir Thomas Cavendish's ship Desire when he arrived there at Christmas time in 1586. This Elizabethan explorer was the second English circumnavigator of the globe.

R. Jan.ʸ 19/35

Beagle
Valparaiso
26. Sept.ʳ /34

(19)

Dear Captain Beaufort

Will you allow the accompanying letters to be forwarded.

They are duplicates of some letters on business – which I have sent with your

your letters – by the Samarang.

Perhaps these duplicates sent by a Merchant man may arrive first.

I am ever

Your's most sincerely and Respectfully –

Robt. FitzRoy

My Schooner is sold –

Our painting man

Mr. Martens is gone.

The Charts are progressing slowly – They are not ready to send away yet – I am in the dumps. It is heavy work – all work and no play – like your office – something – though not half so bad. probably!

God bless you –

admiring himself in a looking glass; and a merry-faced little Indian boy from the Rio Negro, whom we had for some months on board, soon perceived this, and used to mock him: Jemmy, who was always rather jealous of the attention paid to this little boy, did not at all like this, and used to say, with rather a contemptuous twist of his head, 'Too much skylark'.

…Fuegia Basket was a nice, modest, reserved young girl, with a rather pleasing but sometimes sullen expression, and very quick in learning anything, especially languages. This she showed in picking up some Portuguese, and Spanish, when left on shore for only a short time at Rio de Janeiro and Monte Video, and in her knowledge of English. York Minster was very jealous of any attention paid to her; for it was clear he determined to marry her as soon as they were settled on shore.

As *Beagle's* captain FitzRoy was instrumental in selecting his officers and crew. It is a testimony to his prowess as a commander that so many of his former shipmates sailed with him again. Lieutenant Sulivan, and John Lort Stokes, the mate and assistant surveyor, both served on the first *Beagle* expedition. Philip Gidley King, midshipman, was also on the earlier expedition but on the support vessel *Adventure*. The surgeon, Benjamin Bynoe, was another *Beagle* veteran. The nine supernumeraries onboard, including the Fuegians, increased the total number of people sailing on the *Beagle* from sixty-five to seventy-four.

FitzRoy knew how to motivate his men and whenever possible he arranged sports, games and, occasionally, good-natured skylarking. He actively

Woollya in Tierra del Fuego, engraving after Robert FitzRoy from the *Narrative* (1839). On 23 January 1833 the three Fuegians and the Reverend Richard Matthews were settled in Tierra del Fuego, at Woollya, beside Murray Narrows and close to the Beagle Channel (discovered by FitzRoy during *Beagle*'s first expedition).

encouraged and participated in the crossing-the-line ceremony when on 17 February 1832 the *Beagle* crossed the equator. He had read and recorded the words of the Russian explorer Otto von Kotzebue (1787–1846), who had completed a circumnavigation in 1823–6.

FitzRoy noted Kotzebue's words in his *Narrative*:

> *These sports, while they serve to keep up the spirits of the men, and make them forget the difficulties they have to go through, produce also the most beneficial influence upon their health; a cheerful man being much more capable of resisting a fit of sickness than a melancholy one. It is the duty of commanders to use every innocent means of maintaining this temper in their crews; for, in long voyages, when they are several months together wandering on an element not destined by nature for the residence of man, without enjoying even occasionally the recreations of the land, the mind naturally tends to melancholy, which of itself pays the foundation of many diseases, and sometimes even of insanity. Diversion is often the best medicine, and used as a preservative, seldom fails of its effect.*

Beagle's captain had a temper that was familiar to his officers and crew and of concern to FitzRoy himself. He was well aware he needed to keep it in check. In his *Autobiography* Darwin recalled that 'FitzRoy's temper was a most unfortunate one. This was shown not only by passion but by fits of long-continued moroseness against those who had offended him'.

Conrad Martens, who joined *Beagle* as FitzRoy's artist after Augustus Earle left the ship due to illness, was prematurely let go by FitzRoy before the end of the voyage. In a letter to Darwin from Sydney at the end of January 1862 his postscript stated: 'I wonder whether the Admiral "what is now" [is well]. I should like to send my kind regards, if you should see him, but don't if you don't like; coffee without sugar! – you remember.' When the junior officers relieved each other from their posts they would often enquire as to 'whether much hot coffee had been served out'. 'Hot coffee' or 'hot tea' were metaphors used to enquire about their captain's temperament. The mood of the working day, or night, would be determined by how much had been 'poured'.

Darwin famously fell out with FitzRoy due to a difference of opinion over slavery during their stay in Brazil. Darwin's family, notably grandfathers Erasmus Darwin and Josiah Wedgwood, were high-profile activists against slavery. Josiah created a ceramic plaque portraying a kneeling slave in chains, with the phrase coined by himself – 'Am I not a man and a brother?' – that was adopted by the anti-slavery movement. If Darwin had been an officer or seaman he would certainly have received punishment for daring to challenge *Beagle*'s captain. This was a clash, fortunately of a temporary nature, that is a telling story.

According to Darwin, FitzRoy had

> …*just visited a great slave-owner, who had called up many of his slaves and asked them…whether they wished to be free, and all answered 'No'. I then asked him, perhaps with a sneer, whether he thought that the answers of slaves in the presence of their master was worth anything. This made him excessively angry, and he said that as I doubted his word, we could not live any longer together. I thought that I should have been compelled to leave the ship; but as soon as the news spread, which it did quickly, as the captain sent for his first lieutenant to assuage his anger by abusing me, I was deeply gratified by receiving an invitation from all the gun-room officers to mess with them. But after a few hours FitzRoy showed his usual magnanimity by sending an officer to me with an apology and a request that I would continue to live with him.*

Darwin's *Autobiography* is liberally peppered with references to FitzRoy, which include:

> *He was extremely kind to me, but was a man very difficult to live with on the intimate terms which necessarily followed our messing by ourselves in the same cabin. We had several quarrels; for when out of temper he was utterly unreasonable.… The difficulty of living on good terms with a Captain of a Man-of-War is much increased by its being almost mutinous to answer him as one would answer anyone else.*

In terms of discipline, officers and seamen had the expectation that captains could and would mete out punishments as and when required. *Beagle* was delayed in sailing from Devonport at the beginning of December 1831 by poor weather, eventually sailing after Christmas on 27 December. Darwin recorded in his Diary on Christmas Day that 'there was not a sober man in the ship. King is obliged to perform duty of sentry, the last sentinel came staggering below declaring he would no longer stand on duty, whereupon he

is now in irons getting sober as fast as he can'. Darwin believed that *Beagle* had failed to sail from England on 26 December because of the ill effects of the drunken festivities.

Darwin was disturbed by the punishments ordered once they were at sea but realized they were a necessary part of managing men and keeping discipline afloat. For drunkenness, fighting, breaking leave, quarrelling, insolence and neglect of duty FitzRoy 'disrated' (demoted) several men and placed several in irons (chains). FitzRoy's logbook for 28 December (National Archives, Kew) records that James Phipps received forty-four lashes for breaking his leave, drunkenness and insolence. David Russel was given thirty-four lashes for breaking his leave and disobedience. Elias Davis received thirty-one lashes for neglect of duty and John Bruce was given twenty-five lashes for drunkenness, quarrelling and insolence. But this was unusual, and *Beagle*'s voyage was notable for the relatively few punishments carried out.

In his *Narrative* FitzRoy explained his views on maintaining shipboard discipline: 'Hating, abhorring corporal punishment, I am nevertheless fully aware that there are too many coarse natures which cannot be restrained without it, not to have a thorough conviction that it could only be dispensed with, by sacrificing a great deal of discipline and consequent efficiency'.

FitzRoy believed that:
> *To the executive officers of a ship it is always a most satisfactory*

feeling, independent of other thoughts, to be fairly at sea, and away from the scenes of irregularity which so often take place in ports. Those scenes, however, are now much less offensive, and the sailor is far less heedless than he was formerly, if we may take Fielding's description as authority. That humorous sensible author says, in one of the most entertaining accounts of a voyage ever written [Journal of a Voyage to Lisbon (1755)], 'To say the truth, from what I observed in the behaviour of the sailors in this voyage, and comparing it with what I have formerly seen of them, at sea, and on shore, I am convinced that on land there is nothing more idle and dissolute; but, in their own element, there are no persons, near the level of their degree, who live in the constant practice of half so many good qualities.' [Henry Fielding was an English novelist and dramatist best known for his comic novel **Tom Jones** (1749).]

'Salvage of Stores and Treasure from HMS *Thetis* at Cape Frio, Brazil', oil painting by John Christian Schetky, one of a pair of canvases painted in 1833. The ship sank on 4 December 1830 after crashing into the rocks at the base of Cabo Frio (now Thetis Cove), north of Rio de Janeiro, because the captain relied on dead reckoning rather than taking soundings. Among its valuable cargo were silver bars, plates, coins and some gold. Diving bells were constructed out of water tanks and used to recover part of the treasure.

Later in the voyage after the Admiralty stressed that they would not fund additional survey or support vessels to assist the expedition, FitzRoy's state of mind took a turn for the worse. He clearly felt badly let down by their decision. His letter to Beaufort towards the end of September 1834 reveals his troubled mind:

> …I am in the dumps. It is heavy work – all work and no play – like your Office…. Troubles and difficulties harass and oppress me so much that I find it impossible either to say or do what I wish. Excuse me then I beg of you if my letters are at present short and unsatisfac-

tory – Having been obliged to sell my Schooner, and crowd everything again on board the Beagle – Disappointment with respect to Mr. Stokes – also the acting Surgeon – and the acting Boatswain [FitzRoy's attempts to obtain promotion for these men had failed] – Continual hard work – and heavy expense – These and many other things have made me ill and very unhappy.

FitzRoy was confronting his greatest challenge. He desperately needed the support of his officers and crew, and they did not let him down. Darwin recorded an incident in the summer of 1834 in his *Autobiography* that suggests that all was not well with *Beagle*'s captain:

Poor FitzRoy was sadly overworked and in very low spirits; he complained bitterly to me that he must give a great party to all the inhabitants of the place. I remonstrated and said that I could see no such necessity on his part under the circumstances. He then burst out into a fury, declaring that I was the sort of man who would receive any favours and make no return. I got up and left the cabin without saying a word, and returned to Conception [Darwin's memory was playing tricks, it was in fact Valparaiso] where I was then lodging. After a few days I came back to the ship and was received by the Captain as cordially as ever, for the storm had by that time quite blown over. The first Lieutenant [Wickham], however, said to me: 'Confound you, philosopher [one of Darwin's nicknames], I wish you would not quarrel with the skipper; the day you left the ship I was dead-tired (the ship was refitting) and he kept me walking the deck till midnight abusing you all the time.

Gradually FitzRoy became more withdrawn and ceased to actively run the ship. He confided to the surgeon, Benjamin Bynoe, his family history and the possibility that his uncle's suicide could have a bearing on his current mental anguish. He wanted to resign his command and hand it over to his second-in-command, John Clements Wickham.

In a letter to Fanny from Valparaiso on 6 November 1834, he refers to his state of mind, what he described as his 'blue devils':

I am so surrounded with troubles and difficulties of sorry kind that I can only send a short and very stupid letter.... My brains are more confused even than they used to be in London.... Alas how many that were once disposed to be kind to me – my own conduct has estranged. This Survey has indeed done me more harm in every way than it is easy to believe – I have lost by it health,– time, – money, – and friends'.

Darwin also wrote a revealing letter to his sister Catherine from Valparaiso on 8 November 1834:

My last letter was rather a gloomy one, for I was not very well when I wrote it – Now everything is as bright as sunshine. I am quite well again after being a second time in bed for a fortnight. Capt FitzRoy very generously has delayed the Ship 10 days on my account & without at the time telling me for what reason.– We have had some strange proceedings on board the Beagle, but which have ended most capitally for all hands.– Capt FitzRoy has for the last two months, been working extremely hard & at the same time constantly annoyed by interruptions from officers of other ships: the selling the Schooner & its consequences were very vexatious: the cold manner the Admiralty (solely I believe because he is a Tory) have treated him, & a thousand other &c &c has made him very thin & unwell, This was accompanied by a morbid depression of spirits, & a loss of all decision & resolution. The Captain was afraid that his mind was becoming deranged (being aware of his hereditary predisposition). All that Bynoe could say, that it was merely the effect of bodily health & exhaustion after such application, would not do; he invalided & Wickham was appointed to the command. By the instructions Wickham could only finish the survey of the Southern part & would then have been obliged to return direct to England.– The grief on board the Beagle about the Captains decision was universal & deeply felt.– One great source of his annoyment, was the feeling it impossible to fulfil the whole instructions; from his state of mind, it never occurred to him, that the very instructions order him to do as much of West coast, as he has time for & then proceed across the Pacific. Wickham (very disinterestedly, giving up his own promotion) urged this most strongly, stating that when he took the command, nothing should induce him to go to T. del Fuego again; & then asked the Captain, what would be gained by his resignation. Why not do the more useful part & return, as commanded by the Pacific. The Captain, at last, to every ones joy consented & the resignation was withdrawn.

John Clements Wickham was a loyal supporter of FitzRoy. He was never tempted to take advantage of his captain's descent into melancholy. The

Island of Moorea, drawing by Conrad Martens. All the officers and crew longed to escape the arduous conditions and cold climate of South America. Darwin yearned for the warmer climes and exotic sights of the South Pacific. The *Beagle* expedition arrived at Tahiti on 15 November 1835: the island of Moorea is situated about 11 miles to the west of Tahiti. A watercolour development of this drawing (dated 28 January 1836) was sold to FitzRoy when he visited Martens in Australia and was engraved for the *Narrative* (1839).

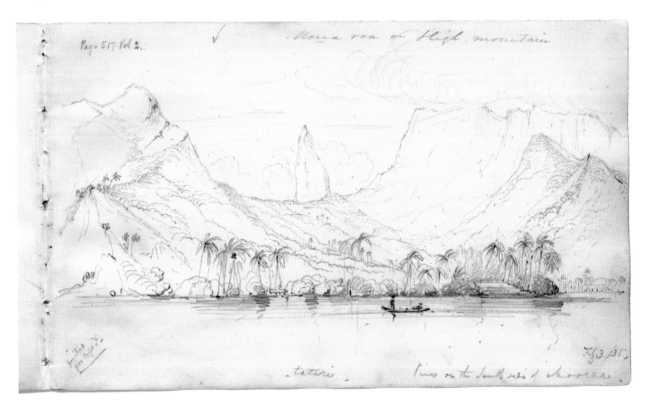

Bay of Islands, New Zealand, watercolour by Augustus Earle. On 21 December 1835 the *Beagle* anchored in the Bay of Islands at the north-eastern tip of North Island. Darwin's great-grandson, Richard Darwin Keynes, described the scene as 'distinctly less than encouraging in comparison with their joyful and boisterous welcome at Tahiti'.

Admiralty Instructions were clear on what FitzRoy should have done in this situation:

> *In the event of any unfortunate accident happening to you yourself, the officer on whom the command of the Beagle may in consequence devolve, is hereby required and directed to complete as far as in him lies, that part of the survey on which the vessel may be then engaged, but not to proceed to a new step in the voyage; as, for instance, if at that time carrying on the coast survey on the western side of south America, he is not to cross the Pacific, but to return to England by Rio de Janeiro and the Atlantic.*

To Wickham's credit he refused to take command from FitzRoy. He was adamant that he was not going to accept FitzRoy's order. The first lieutenant's refusal was insubordinate but it prompted FitzRoy to rethink his responsibilities and he decided that there was nothing to be gained by renouncing command and withdrew his resignation. If Wickham had accepted, FitzRoy would have been court-martialled upon their return. Wickham saved his captain's career and the integrity and success of the *Beagle* expedition.

FitzRoy's leadership skills were stretched to the limit, but he personally benefited from a loyal and supportive team of professional officers and seamen. FitzRoy was well aware of his personal failings and fortunately he was able, with the help of his men and messmate, to avoid the fate that had befallen *Beagle*'s previous commander, Pringle Stokes.

FitzRoy fulfilled his promise to return the Fuegians to their home-waters; however, the civilizing experiment had failed and they reverted to their native manners, customs and habits. Although they were not all from the same tribe, at the request of York Minster, they were all settled together at Woollya in Tierra del Fuego towards the end of January 1833. Darwin wrote of the items transported and unpacked for the settlement:

> *The choice of articles showed the most culpable folly and negligence. Wine glasses, butter-bolts, tea trays, soup turins, mahogany dressing case, fine white linen, beaver hats and an endless variety of similar things, show how little was thought about the country where they were going to. The means absolutely wasted on such things would have purchased an immense stock of really useful articles.*

Darwin was sad to see the Fuegians leave the ship and prophetically recorded in his Diary that, 'They have far too much sense not to see the vast superiority of civilized over uncivilized habits, yet I am afraid to the latter they must return'.

Later FitzRoy discovered that York Minster had robbed Jemmy Button of his possessions and settled elsewhere with Fuegia Basket. He revisited the settlement several times during *Beagle*'s expedition, but within a year it had been abandoned.

On 5 March 1834 FitzRoy inspected the camp to find the wooden wigrams still standing but empty and the vegetable garden trampled. But he did

View of Port Jackson, Australia, watercolour by Conrad Martens. Early in the morning of 12 January 1836 a light air carried the *Beagle* towards the entrance of Port Jackson. The ship's chronometers were used to accurately fix the longitude position of the observatory at Parramatta.

observe Jemmy Button through a telescope heading towards the *Beagle* in a canoe with his brother, Tommy Button. FitzRoy did not recognize him at first. Jemmy was naked and, according to FitzRoy, 'wretchedly thin'. He dined in the captain's cabin and had not forgotten his English lessons or table manners. His wife, family and friends had also learned some English words, and called him Jemmy Button. These facts were hardly consolation for taking the Fuegians to England in the first place.

FitzRoy must have constantly pondered whether his actions had been justified and worthwhile. When the *Beagle* arrived in New Zealand in late December 1835, he discussed his hopes for the Fuegians at length with the missionaries who were based there. They were very encouraging and reassured FitzRoy that, far from being a failure, what he had attempted was a significant first step.

The Reverend Richard Matthews, the missionary who had accompanied the Fuegians, tried his best to settle in Tierra del Fuego. However, it proved a clash of cultures that, had he stayed, would have inevitably resulted in his untimely death: Matthews had been physically threatened. FitzRoy thought he was 'rather too young, and less experienced than might have been wished, his character and conduct had been such as to give very fair grounds for anticipating that he would, at least, sincerely endeavour to do his utmost in a situation so difficult and trying as that for which he volunteered'. Matthews eventually settled in New Zealand where his brother was also resident as a missionary.

Inspired by *Beagle*'s passing Cape Frio on Brazil's south-east Atlantic coast on 18 March 1832, FitzRoy would later include an account in the *Narrative* of the loss of the frigate HMS *Thetis* off that Cape in December 1830. He and Sulivan had both served aboard this ship. The vessel sank close to port with a large loss of life and vast treasure onboard. John Christian Schetky, FitzRoy's College drawing professor, painted two paintings in 1833 relating to the salvage of her stores and treasure. They are now in the National Maritime Museum at Greenwich. FitzRoy's discussion of how the ship foundered reveals a man driven by a sense of duty. He compared the risks taken by Royal Navy vessels to those of merchant ships:

> *As in the case of the Thetis, an English man-of-war may incur risk in consequence of a praiseworthy zeal to avoid delaying in port, as a merchant-ship would probably be obliged to do, from her being unable to beat out against an adverse wind, and, like that frigate, may be the first to prove the existence of an unsuspecting danger.*
>
> *Those who never run any risk; who sail only when the wind is fair; who heave to when approaching land, though perhaps a day's sail distant; and who even delay the performance of urgent duties until they can be done easily and quite safely; are doubtless, extremely prudent persons:- but rather unlike officers whose names will never be forgotten while England has a navy.*

After the completion of the core survey work and FitzRoy's decision not to resign his command, his spirits improved once the *Beagle* left South

American waters. But he still had a lot on his mind. FitzRoy recorded details of the Great Earthquake of 1835 at the city of Concepcion in Chile. The stopovers to establish the precise navigational positions using the chronometers were achieved. On Tahiti he was impressed by the success of the missionaries in civilizing the native population. He had extraordinary meetings with a genuine member of the Tahitian royal family, Queen Pomare, but was highly suspicious and wary of an English-born pseudo-royal pretender who was temporarily resident there, and pretentiously called himself Baron Charles Philippe Hippolytus de Thierry. Baron de Thierry was also heading for New Zealand, apparently to claim his sovereign rights.

In Sydney, Australia, FitzRoy paid a visit to Conrad Martens, who had settled there after leaving the *Beagle*. He commissioned from the artist a watercolour view of the island of Moorea, near Tahiti, presumably as a souvenir of the voyage. Paying him two guineas, FitzRoy later commissioned Martens to work up several watercolours from his *Beagle* sketchbooks. He also acquired additional pictorial material that was engraved for volumes I and II of the *Narrative*.

FitzRoy was welcomed by *Beagle's* first expedition commander Phillip Parker King, who was by this time resident in Sydney. King's letter to Beaufort reveals that he clearly thought very highly of FitzRoy and was concerned about the effects upon him of the last four years of labour:

> *I regret to say he has suffered very much and is yet suffering from ill*
> *health – he has had a very severe shake to his constitution which a*
> *little rest in England will I hope restore for he is an excellent fellow*
> *and will I am satisfied yet be a shining ornament to our service.*

In terms of vivid descriptions and insights relating to the later stages of the voyage it is Darwin who provided the most impressive material. The stopovers were completed as quickly as possible. Almost everyone longed to return home.

After more than four years and nine months, FitzRoy brought *Beagle* safely to English shores, anchoring at Falmouth on 2 October 1836. He had circumnavigated the world and successfully completed a large part of his official instructions to undertake survey work and exploration in and around South America. Many of his charts were still in use until the end of the Second World War.

Beagle's captain could be arrogant and he could also be humble. Writing to his sister he expressed humility: 'Our voyage has been more successful than I had any right to anticipate. We have been extremely fortunate in all ways'.

Nelson's celebrated signal that all could see around the rim of *Beagle's* wheel sums up the spirit of FitzRoy's achievement. FitzRoy was cut from the same cloth as his celebrated forebear. As *Beagle's* commander he motivated, cared for and inspired loyalty in his men. He was a man of action who took calculated risks and was not afraid to defy Admiralty orders to get the job done. But above all FitzRoy had done his duty.

Chapter 3
FitzRoy's Officers & Crew

'I like the officers much more than I did at first, especially Wickham & young King & Stokes & indeed all of them. The Captain continues steadily very kind & does everything in his power to assist me.' – Charles Darwin in a letter to his father, begun 8 February 1832

HMS *Beagle* – Officers and Crew

John Clements Wickham, Lieutenant
Bartholomew James Sulivan, Lieutenant
Edward Main Chaffers, Master
Robert McCormick, Surgeon
George Rowlett, Purser
Alexander Derbishire, Mate
Peter Benson Stewart, Mate
John Lort Stokes, Mate and Assistant Surveyor
Benjamin Bynoe, Assistant Surgeon
Arthur Mellersh, Midshipman
Philip Gidley King, Midshipman
Alexander Burns Usborne, Master's Assistant
(and temporary assistant to Conrad Martens)
Charles Musters, Volunteer First Class
Jonathan May, Carpenter
Edward H. Hellyer, Clerk
Acting boatswain: sergeant of marines and seven privates;
thirty-four seamen and six boys.

On the List of supernumeraries were –

Charles Darwin, Naturalist
Augustus Earle, Draftsman
Conrad Martens, Draftsman and Artist (Earle's replacement)
George James Stebbing, Instrument Maker
Richard Matthews and three Fuegians: my own steward:
and Mr. Darwin's servant
(Source: FitzRoy's *Narrative*, 1839)

Everyone aboard HMS *Beagle* had a specific working role, with the exception of the three Fuegians. The missionary Reverend Richard Matthews, who was travelling to assist and settle the Fuegians, was also signed onboard as a supernumerary, and probably assisted with divine services afloat and ministered to the men. FitzRoy had hoped that two missionaries would travel with him to establish a Christian missionary in Tierra del Fuego, but only one could be found who was up for the challenge.

Beagle was a commissioned naval vessel with specific instructions to carry out survey work in all weather conditions. All hands pulled together to this end. The captain was the highest ranking officer but the ship could not operate effectively without a cook, or a carpenter.

Darwin reveals in his Diary the names and roles of many other men onboard who are not mentioned by FitzRoy or the other officers. Writing on 24 July 1832, Darwin,

> *…procured this evening a Watch-bill & as most likely our crew will for rest of the voyage remain the same.– I will copy it.– Boatswains mates. J. Smith & W. Williams: — Quarter-Masters. J. Peterson. White. Bennett. Henderson: — Forecastle Men, J. Davis, (gunmen Heard, Bosworthick (rope maker); Tanner; Harper (sail maker); Wills (armourer); — Fore top-men, Evans; Rensfrey; Door. Wright; Robinson; MacCurdy; Hare; Clarke; — Main top-men Phipps; J. Blight; Moore; Hughes; Johns B.; Sloane; Chadwick; Johns; Williams; Blight, B.; Childs; — Carpenters crew, Rogers; Rowe; J. May; James; — Idlers, Stebbing (instrument mender); Ash, gunroom steward; Fuller, Captains do; R. Davis, boy do; Matthews, missionary; E. Davis, Officers cook; G Phillips, ships cook; Lester, cooper; Covington, fiddler & boy to Poop-cabin; Billet, gunroom-boy; Royal Marines, — Beareley, sergeant; William, Jones, Burgess, Bute, Doyle, Martin, Middleton, Prior (midshipmen steward); — Boatswain, Mr Sorrell; Carpenter, Mr May.*

The list shows that there were, in fact, two cooks, one for the officers and another for the crew, as well as a carpenters' crew led by Jonathan May, and seven midshipmen – Stewart, Usborne, Johnson, Stokes, Mellersh, King and Forsyth.

Darwin also listed the officers and, according to his calculations (which differed from FitzRoy's), there were '(including Earl, the Fuegians & myself) 76 souls on board the Beagle'.

Some of the officers and crew were veterans of the first *Beagle* expedition. Writing in the *Narrative* FitzRoy noted:

> *Many of the crew had sailed with me in the previous voyage of the Beagle; and there were a few officers, as well as some marines and seamen, who had served in the Beagle, or Adventure, during the whole of the former voyage. These determined admirers of Tierra del Fuego were, Lieutenant Wickham, Mr. Bynoe, Mr. Stokes, Mr. Mellersh, and Mr. King; the boatswain, carpenter, and sergeant; four private marines, my coxswain, and some seamen.*

A number of them participated in all three of *Beagle*'s expeditions. They included the surgeon Benjamin Bynoe and the naval officer John Lort Stokes. These men were familiar with the risks associated with going to sea, although the dangers were considerably less than those faced by men on active service in wartime. Even so, the prospect of death from disease, drowning and injury was real enough.

Early in the ship's second expedition midshipman Charles Musters (d. May 1832) and two seamen died of yellow fever.

Darwin recorded in his *Diary* on 4 June 1832 while resident in Rio de Janeiro that Philip Gidley King had brought the calamitous news of

> ...*the death of three of our ship-mates – They were of the Macau party who were ill with fever when the Beagle sailed from Rio – 1st Morgan, an extraordinary powerful man & excellent seaman; he was a very brave man & had performed some curious feats, he put a whole party of Portuguese to flight, who had molested the party; he pitched an armed sentinel into the sea at St Jago; & formerly he was one of the boarders in that most gallant action against the Slaver the Black Joke. – 2d Boy Jones one of the most promising boys in the ship & had been promised but the day before his illness, promotion. – These were the only two of the sailors who were with the Cutter, & picked for their excellence. – And lastly, poor little Musters; who three days before his illness heard of his Mothers death. Morgan was taken ill 4 days after arriving on board & died near the Abrolhos, where he was lowered into the sea after divisions on Sunday – for several days he was violently delirious & talked about the party. – Boy Jones died two days after arriving at Bahia, & Musters two days after that.*

George Rowlett, the purser, probably died of phthisis. Aged 38 he was the oldest seaman onboard and some thought he had died of old age. On 27 June 1834 he was buried at sea when the *Beagle* was sailing between the island of Chiloe off the coast of Chile and Valparaiso. One of the youngest shipmates, Edward Hellyer accidentally drowned in the Falkland Islands during one of the many hunting trips.

Hellyer was FitzRoy's personal clerk. Writing to Captain Beaufort from Montevideo on 26 October 1833, FitzRoy recounted:

> *In Berkeley Sound, not half a mile from the Beagle [in March 1833], he shot a curious bird, and anxious to get it out of the water, he stripped and swam for it; the seaweed caught and entangled his legs, and the tide rose over his head. A melancholy end for one of the worthiest young men I ever knew.*

At the end of the Napoleonic Wars, the number of British naval ships had, according to British naval historian Christopher Lloyd (former Professor of History, Royal Naval College, Greenwich), been rapidly run down from 715 in 1815 to 134 in 1820. Manpower was also cut dramatically during the same timeframe from 140,000 to 23,000. There were plenty of men out of work and there was no need to impress men into the survey ships.

What were the incentives for putting your life on the line for the Royal Navy during this period and did the men give a fig for their King, Queen or country? Within Victorian Britain, and the decades leading up to it, a sense of duty was deemed by the officers, and arguably the majority of the men, as being of the utmost importance. Admiral Nelson's famous signal at Trafalgar in 1805, 'England Expects That Every Man Will Do His Duty', and the decisive actions of the men behind them, had an influential effect not only on the Royal Navy, but on civilian morale, too. Today it is difficult for most people to recognize and acknowledge the significance of duty. For the majority of modern citizens, with the notable exception of our fighting forces and emergency services, it is the family that comes first. Queen and country might follow somewhere behind.

In FitzRoy's preface to the first volume of the *Narrative* there is no doubt that duty is of paramount importance to him: 'In this work, the result of nine years' voyaging, partly on coasts little known, an attempt has been made to combine giving general information with the paramount object – that of fulfilling a duty to the Admiralty, for the benefit of Seamen'. He concludes: 'I beg to remind the reader, that the… publication arises solely from a sense of duty'.

Ironically when 'England Expects…' was hoisted immediately before the Battle of Trafalgar by Nelson's signal officer John Pasco, the reaction of Cuthbert Collingwood, Nelson's second-in-command, was one of bemusement. He is traditionally believed to have said: 'What is Nelson signalling about. We all know what we have to do?' Some sailors allegedly grumbled that they had always done their duty. Interestingly, John Pasco's youngest son, Plymouth-born Crawford Atchison Denman Pasco (1818–1898), transferred to HMS *Beagle* during her third survey voyage, from HMS *Britomart* during her expedition of 1837–43, when the two vessels came into contact at Port Essington in north-western Australia.

For some, the demanding conditions afloat were on a par with and perhaps in some instances superior to what would have been their rural and urban lives on land. Of course the men were subjected to naval codes of conduct, and the marines were onboard to maintain order and discipline. But certainly none of *Beagle*'s captains could be accused of being an out-and-out bully or tyrannical commander.

FitzRoy certainly had a 'hot temper' but had strong leadership qualities and a commanding presence on deck. Darwin described him as someone he could
> …*fancy as a Napoleon or a Nelson.… It is very amusing to see all*
> *hands hauling at a rope they not supposing him on deck & then*
> *observe the effect, when he utters a syllable: it is like a string of*
> *dray horses, when the waggoner gives one of his awful smacks.*

Some of FitzRoy's men were a little fearful of him but he was not overly excessive in ordering punishments. An Admiralty order of 1830 urged 'a safe forbearance' from captains, making more than two-dozen lashes without a trial illegal. However, flogging died out slowly and was only suspended during peacetime in 1871. FitzRoy's approach earned him the respect and loyalty of the men, and sometimes admiration, too.

FitzRoy's *Additional Orders* (1–59) dated 4 September 1832 were drawn up and issued during the *Beagle* expedition (a copy is now in the Caird Library, National Maritime Museum, Greenwich) and reveal the lengths he went to tell his officers what to expect from him, how to handle shipboard discipline and set a good example. Order No. 24 stated: 'The officers may rely on my supporting them in their duty…but they must be careful not to let anger induce them to take the law into their own hands, or, to use bad language'.

Order No. 26 reveals the intelligence of *Beagle*'s captain. FitzRoy was determined that shipboard orders should be delivered to his men in an effective and efficient manner.

> *When an officer's voice is made too common or when all orders are given in the same tone to a man at the mast head as well as to a man at the foot of the mast respect for and attention to that voice is much lessened. Certainty of Punishment without severity is a maxim which will be adhered to in this vessel.*

Good captains motivate their men. Darwin also described FitzRoy as his 'beau ideal of a captain'. From Darwin's Diary we learn that at Port Desire on Christmas Day (1833):

> *After dining in the Gun-room, the officers & almost every man in the ship went on shore. The captain distributed prizes to the best runners, leapers and wrestlers. These Olympic games were very*

'Slinging the Monkey' at Port Desire on Christmas Day 1833, graphite and water-colour by Conrad Martens. The naval game 'slinging the monkey', played here, involved a seaman slung upside down from a tripod. If the seaman with a stick was able to hit one of the surrounding seamen, busy taunting and 'beating' the 'monkey', that man would take his place. FitzRoy provided Martens with nautical advice by inscribing the picture, 'Note Mainmast of Beagle a little father aft, Miz. [Mizzen] Mast to rake more'.

amusing; it was quite delightful to see what school-boy eagerness the
seamen enjoyed them: old men with long beards & young men
without any were playing like so many children. Certainly a much
better way of passing Christmas day than the usual one, of every
seaman getting as drunk as he possibly can.

The Royal Navy had a vested interest in keeping their men alive. The accommodation was adequate, albeit cramped. Regular food and drink were provided throughout the day. The daily allowance of rations per man comprised: 'Bread 16 oz, Meat (beef or pork) 16 oz, Vegetables 8 oz, Cocoa 1 oz, Tea ¼ oz, Sugar 1½ oz, Beer 1 gallon (wine was substituted on foreign stations), Flour 6 oz and every week ½ lb currants and ½ lb suet or oatmeal was added' (taken from *Medicine And The Navy*, Volume IV, 1815–1900). Christopher Lloyd and Jack Coulter claim that, 'On paper it was in quantity, if not quality, a superior diet to that enjoyed by most members of the labouring classes on shore, because during the first half of the nineteenth century the conditions of these classes was probably worse than at any other time in their history. FitzRoy wanted the best for his men. His Additional Order No. 27 reveals that, 'The ships company are always to be allowed three quarters of an hour for breakfast, one hour and a half for dinner and three quarters of an hour for supper when in harbour'. At sea they were allowed the corresponding times of thirty minutes, one hour and half an hour.

On most days Darwin took his meals with FitzRoy in the captain's cabin. He described his daily dining routine and the quality of the food:

We breakfast at 8 o'clock. The invariable maxim is to throw away
all politeness:- that is, never to wait for each other, and bolt off the
minute one has done eating, etc.... At one we dine. You shore-going
people are lamentably mistaken about the manner of living on board.
We have never yet (nor shall we) dined off salt meat. Rice and Pea and
Calavanses [a type of pulse] are excellent vegetables, and with good
bread – who could want more? ... At 5 we have tea.

Darwin also observed that 'nothing but water comes on the table'.

The popularity of grog (rum diluted with water at a ratio of 1:2) sometimes presented a serious problem aboard naval ships with outbreaks of severe drunkenness, especially at Christmas time. In 1825 the daily allowance was reduced from a half-pint to a quarter-pint at dinner and supper. In tropical climes the ship's surgeon would sometimes mix the daily issue with lemon juice to stave off scurvy.

The standard pay for sailors certainly did not reflect in modern terms their hazardous occupation or working hours, and it was paid after all the various deductions for their keep were taken into account.

John Lort Stokes' personal signed copy of the *Admiralty Regulations* for 1826 (full title *Regulations established by The King in Council and Instructions, issued by the The Lords Commissioners of the Admiralty relating to His*

Majesty's Service at Sea) is now in the Caird Library of the National Maritime Museum. It is inscribed September 1831 and Stokes would have taken it with him on the *Beagle*. Among many other rules, regulations and guidelines for the officers and seamen, the publication also states the breakdown of naval wages. An ordinary seaman was paid £1 6d per month, while the more experienced able seaman received £1 14s, the bosun £4 14s, mate £2 18s, cook £2 12s and midshipman £2 8s. The senior surgeon was paid £12 5s 4d, while his assistant received £9 4s, and the ship's master, who was in charge of navigational matters, was paid £7 13s.

Senior officers fared better: the junior lieutenants received £9 4s, which increased incrementally according to their number of years of service, while the commander or captain could expect around £23 4d. In fact the official muster books of the *Beagle* reveal higher amounts paid by the purser to officers and crewmen. Between 1 January and 29 February 1832 the muster book reveals 'the Bills drawn and Monthly Allowance paid' to John Clements Wickham of £53 19s 2d, to Sulivan £55 0s 2d, and to FitzRoy £103 11s 6d, while the ship's carpenter, Jonathan May, received a two-month payment of £7 17s 4d. There was little chance of supplementing naval pay with prize money in the capture of enemy ships.

A sense of adventure, camaraderie and companionship were all additional motivational factors for a life at sea. You could not exactly describe the presence of a surgeon, and his assistant, as a medical benefit, but they were on hand. They were never short of patients.

During FitzRoy's command of the *Beagle* she carried a wide range of supplies to combat scurvy, and to prevent and cure other illnesses. She had 'various antiscorbitants – such as pickles, dried apples, and lemon juice – of the best quality, …also a very large quantity of Kilner and Moorsom's preserved meat, vegetables, and soup'. An ample supply of antiseptics was carried, but in un-chartered waters it was inevitable that parasitical infections and diseases for which the men had no natural defence would occur. Even so, *Beagle*'s medical men held an impressively low mortality record.

One of the traditional roles of the surgeon aboard a ship was to collect specimens and examples of flora and fauna during a voyage. But on *Beagle* this role was passed to Darwin. Robert McCormick, *Beagle*'s ship's surgeon, was displeased that Darwin had usurped his position. It is possible that FitzRoy had not fully briefed him before the ship set sail, but McCormick left the ship in April 1832 in a fit of pique after having fallen out with his captain. Darwin was not sad. He was normally so polite and even-tempered, but he remarked, 'my friend the Doctor is an ass'. McCormick was succeeded by the well-regarded and amiable assistant surgeon, Benjamin Bynoe.

FitzRoy expected dedication and hard work from his men, and for those who met his exacting standards he was not hesitant in approaching the Admiralty to stress their achievements. FitzRoy actively tried to advance several of his men's careers by recommending promotion; however, he was not always successful.

In April 1835 FitzRoy himself was promoted to a full Post-Captain by rank. The Admiralty was not aware of his severe depression and his threat of relinquishing his command. The term post-captain is no longer used today in the Royal Navy. It was used to distinguish those who were captains by rank from officers in command of a naval vessel, who were (and still are) addressed as captain regardless of rank. Once an officer had reached the position of post-captain, promotion was then (during this period) strictly based on seniority.

John Clements Wickham by an anonymous photographer. Wickham took command of *Beagle* on her third, and last, survey expedition to Australia.

However, FitzRoy's efforts to gain promotion or reward for John Clements Wickham and John Lort Stokes were passed over. FitzRoy was far from happy. Without Wickham, FitzRoy's career would have been cut short and he would have received no promotion.

John Clements Wickham (1798–1864) was *Beagle*'s first lieutenant. He was the most senior officer after the captain. He had a long naval service: he had served as second lieutenant under Phillip Parker King on the *Adventure* during *Beagle*'s first expedition and would serve as commander and surveyor on the *Beagle*'s third expedition. Born in Leith, Scotland, Wickham followed his father, Lieutenant Samuel Wickham, into the Royal Navy. In 1812 he joined as a midshipman and was promoted to Captain in 1837. He was well liked; however, his protracted career at sea took its toll on his personal health.

Wickham became firm friends with Darwin during the voyage, although they occasionally aired their differences. He used to get annoyed by the mess created when Darwin dumped his fossils and other specimens on the ship's deck. Darwin had discovered fossils in the cliffs at Bahia Blanca in September 1832. Darwin's daughter, Henrietta, recalled:

My Father used to describe how Wickham, the first Lieutenant – a very tidy man who liked to keep the decks so that you could eat your dinner off them – used to say If I had my way, all your d—d mess would be chucked overboard, & you after it old flycatcher.

However, although Wickham was 'always growling at (my father) bringing more dirt on board than any ten men, he is far the most conversible being on board. I do not mean talks the most, for in that respect Sulivan quite bears away the palm'. Darwin thought him 'a glorious fine fellow'.

Fitting for his senior position Wickham took an active part leading survey expeditions on land, and also in the ship's boats. In December 1832 the *Beagle* arrived at the bay of San Blas, south of Bahia Blanca in Argentina, where Wickham reported on the excellent progress on the surveys.

By December 1833 Wickham was placed in command of the *Adventure*, formerly a sealing schooner of 170 tons called the *Unicorn*, which FitzRoy had purchased earlier in the year at Port Louis at the Falkland Islands to facilitate and speed up the survey work. FitzRoy became gravely concerned about Wickham's welfare when he later departed with two other hired vessels, the *Paz* and *Liebre*, on a survey expedition and failed to return.

Luckily Wickham eventually met up again with the *Beagle*. He was given command of another small schooner called the *Constitucion*, which was initially chartered and then acquired by FitzRoy.

Credit must also be given to Wickham (although the veracity of the story has been doubted by some scholars) for his part in the preservation of a remarkable Darwin acquisition – Harriet the Galapagos tortoise. After a short stay in England, Wickham transported the tortoise to Australia, arriving at Brisbane in 1842. Harry, as she was known until an inspection in the 1960s revealed that the old 'fella' was in fact an old girl, was transported to a home in Brisbane City Botanic Gardens. In 1998 she was transferred to Steve Irwin's Australia Zoo on the Gold Coast. In 2005 Irwin announced that Harriet, the giant land tortoise, was celebrating her 175th birthday. DNA tests indicated that she was born around 1830. She had been collected by Darwin in 1835 when she was about the size of a dinner plate. Her body shape revealed that she originated by the island of Santa Cruz (also called Indefatigable Island). Sadly, Harriet passed away peacefully at the Australia Zoo on 23 June 2006.

Ill health forced Wickham to retire from the Royal Navy in 1841, and in the following year he settled in New South Wales, Australia. He was appointed to various official positions, including police magistrate at Moreton Bay, and was actively involved in the colonial government of the new colony of Queensland. But he was frustrated by what he regarded as political indecision, poor judgement, and the quarrelling between his superiors in the governments of New South Wales and Queensland, which also affected his own position. He left Australia and settled in Biarritz in France, where he died.

Darwin checking the speed of a Galapagos tortoise, by Meredith Nugent (1860–1937). Nugent was born in London but emigrated to the west coast of America in the 1930s. He studied in New York, at the Pennsylvania Academy of Fine Arts under Thomas Eakins, and Académie Julian in Paris. He was a popular artist-illustrator for books, journals and magazines, including *Harpers*, *Century* and *Ladies Home Journal*.

Bartholomew James Sulivan (1810–1890) was FitzRoy's second lieutenant and ardent supporter. Like FitzRoy, he trained at the Royal Naval College, Portsmouth, and served with FitzRoy on the first *Beagle* voyage. He was active in the survey work conducted from the ship's boats and accompanied Darwin on shooting expeditions. On 20 February 1832 the *Beagle* arrived at Fernando Noronha, an archipelago around 220 miles off the Brazilian coast. Darwin noted that Sulivan harpooned a large porpoise and that shortly afterwards 'a dozen knives were skinning him for supper'.

Sulivan was a loquacious and well-respected crewmember. He was a fluent Spanish speaker, too, which proved useful in South America. Sulivan had a good sense of humour and adored playing pranks. Sailing towards Rio de Janeiro on 1 April 1832 he shouted out: 'Darwin, did you ever see a Grampus: Bear a hand then'. Apparently Darwin fell for the ruse and rushed to see the imagined animal. There are various interpretations of what a Grampus might be. They include a pilot or killer whale or blunt-nosed dolphin.

Sulivan was also a man of decisive action, who, like Wickham, was not afraid to refute his captain's orders for the sake of the ship's safety. Sulivan's initiative on one occasion arguably saved the *Beagle* from sinking. The story was recounted in *Life and Letters*. Richard Darwin Keynes (Darwin's great-grandson) also summed up the precarious situation that occurred in Tierra del Fuego in January 1833:

Vice-Admiral Sir Bartholomew James Sulivan, portrait from his son's biography (1896). Sulivan and Darwin kept in touch throughout their lives, and Sulivan visited Down House.

> *The Captain always had the ports secured, but that he himself [Sulivan] never liked this order, and told the carpenter always to have a handspike handy for eventualities. Shortly before the three huge waves struck the boat, Sulivan had relinquished the deck to FitzRoy, but on returning up to his waist in water and standing on the bulwark, strenuously driving a handspike against the port, which he eventually burst open. Darwin's recollection appears to confirm that the ports did have to be knocked out in order to right the ship, so that it was evidently Sulivan's countermanding of FitzRoy's standing order that finally saved the day.*

There is no record of FitzRoy chastizing Sulivan for disobeying his standing order of securing the ports (*Beagle*'s gun-ports were in fact fitted with half-ports). He probably admired Sulivan's initiative in breaking one open to ensure that the sea ran off the ship.

Sulivan rose in the Royal Navy to eventually become a vice-admiral. He continued to undertake survey work on the South American Station and the Falkland Islands during 1842–6 in HMS *Philomel*. Sulivan assisted Captain Charles Hotham during a naval campaign in South America in the mid-1840s and was rewarded with promotion. This encouraged Beaufort to send him to assist Admiral Sir Charles Napier (1786–1860), who commanded the British fleet during the Crimean War (1853–6). Sulivan was appointed as a fleet surveying officer in the paddle-vessel *Lightning* (3 guns) from 25 February 1854 to 12 March 1855.

John Lort Stokes, oil portrait by Stephen Pearce (1819–1904). Stokes was a gifted writer and after returning to England following his participation in *Beagle*'s third and final survey expedition to Australia, his two-volume *Discoveries in Australia* was published (London, 1846). He ended his career as admiral on the retired list in 1877.

For a decade he was Marine Adviser to the Board of Trade and was knighted in recognition of his naval services. In this capacity he wrote a pamphlet, dated 5 April 1858, termed 'Proposal for organizing a scientific corps of Naval Officers', in which he made a series of constructive recommendations about encouraging officers into the surveying service not only in peace, but also in hostile times. He noted: 'If the surveying service is to be kept in an effective state, so as to supply the general service with a staff of surveyors in time of war, there must be some inducement held out to officers to adopt that branch of service'.

John Lort Stokes (1812–1885) served on all three of *Beagle*'s survey voyages: as midshipman, mate and assistant surveyor under FitzRoy, and lieutenant and assistant surveyor on the third expedition. Stokes was born in Prendergast, Pembrokeshire in Wales. He was not related to Pringle Stokes. Darwin counted him among his friends, although he made relatively few references to him in his journals and correspondence. During the day Stokes shared Darwin's cabin, which was dominated by the chart table: at night he slept in a bunk in an area under the steps at the entrance to the cabin.

There is a great deal more information available relating to Stokes on *Beagle*'s third survey voyage, because he took over command when Captain John Clements Wickham became ill. Marsden Hordern, a former Royal Australian Naval officer who served in the Second World War, wrote an award-winning biography of Stokes entitled *Mariners are Warned!*, published in 1989. He described Stokes as a 'sober, conscientious and earnest man, [who was prone to] acts of breathtaking foolhardiness'. He escaped drowning several times and experienced close encounters with crocodiles.

Stokes exhibited a flair for leadership and initiative afloat. He had an established track record for survey work at sea but little affinity for land-related matters. His prediction that Port Grey was a non-starter as a base for a colony was misplaced. It thrived, and became one of the earliest successful wheat-growing regions in Western Australia.

Stokes ended an eventful naval career as an admiral on the retired list in 1877. He retired to his estate near Haverfordwest in Pembrokeshire, Wales. He was a lively writer and his two-volume *Discoveries in Australia* was published in 1846. As a testament to his skills, and to those of Wickham too, some of their maps of North Australian waters were still in use during the Second World War. He made a significant contribution to the surveying and charting of the Bass Strait. He also discovered and named the Albert, FitzRoy and Flinders rivers, and Port Darwin, which he named after his friend and shipmate Charles Darwin.

On 30 June 1960 *The Manchester Guardian* newspaper announced that, 'A lifetime's collection of naval relics from John Lort Stokes, of the *Beagle* has been acquired by the National Maritime Museum, Greenwich'. It included logbooks, rough charts, navigational and surveying instruments, as well as uniforms, including the white drill coat uniform with a hole in it made when he was speared by an Aborigine in Australia during *Beagle*'s third survey expedition. Fortunately he survived the injury.

FitzRoy's letter summarizing Stokes's strengths and contribution to the second *Beagle* expedition, sent to the Commissioners of the Admiralty and written before the *Beagle* was paid off at Woolwich (17 November 1836) was first class: 'I know not the man I should prefer to him in a professional way – a surveyor, or in a private capacity as a staunch and sensible friend. His sterling value I have found by long acquaintance and have proved by many a trial'.

Midshipman Philip Gidley King (1817–1904) joined the *Beagle* in 1831. He was born at Parramatta, New South Wales, into a distinguished naval family and therefore it is not surprising that he has been overshadowed, at least in terms of his maritime achievements, by his forebears.

King's grandfather was also called Philip Gidley King (1758–1808). He was born at Launceston in Cornwall, England. On 22 December 1770 he joined the navy as a captain's servant in HMS *Swallow*. Later he was actively involved in Australia's early settlement period and became governor of the colony of New South Wales in 1800–6. As an enlightened man, he encouraged exploration and gathered specimens for Sir Joseph Banks, the President of the Royal Society in London.

It is known that King Junior sailed with his father, Phillip Parker King, on HMS *Adventure* during *Beagle*'s first expedition. During his service on this ship and *Beagle* he engaged in survey work of the South American coast. In 1832 Darwin described the 14-year-old lad as 'the most perfect, pleasant boy I ever met'. They were often in company on land excursions.

Sailors crossing the equatorial line for the first time were subjected to

Detail of a letter from Robert FitzRoy to Capain Francis Beaufort recommending Stokes for promotion. FitzRoy expected a lot from his men but whenever possible he sent letters of recommendation to Beaufort and officials at the Admiralty to advance their career prospects.

a ceremony called crossing the line. This involved mock shaving and ducking in water on deck (sometimes in the sea) with officers and seamen dressed up as King Neptune and Queen Amphitrite, barbers and surgeons, and various other dignitaries. This initiation rite sometimes resulted in a certificate being presented to the victim to exempt the sailor from being subjected to further ceremonies.

King recalled from memory Darwin's reaction to the crossing-the-line ceremony in February 1832:

> *The effect produced on the young naturalist's mind was unmistakably remarkable. His first impression was that the ship's crew from Captain downwards had gone off their heads. 'What fools these sailors make of themselves,' he said as he descended the companion ladder to wait below till he was admitted. The Captain received his godship and Amphitrite his wife with becoming solemnity; Neptune was surrounded by a set of the most ultra-demonical looking beings that could be imagined, stripped to the waist, their naked arms and legs bedaubed with every conceivable colour which the ship's stores could turn out, the orbits of their eyes exaggerated with broad circles of red and yellow pigments. Those demons danced a sort of nautical war dance exulting on the fate awaiting their victims below.*

King also provided an insight into Darwin's nautical abilities:

> *Though Mr Darwin knew little or nothing of nautical matters, on one day [in April 1832] he volunteered his services to the First Lieutenant. The occasion was when the ship first entered Rio Janeiro. It was decided to make a display of smartness in shortening sail before the numerous men-of-war at anchorage under the flags of all nations…. At the order 'Shorten Sail' he was to let go and clap on to any rope he saw was short-handed – this he did and enjoyed the fun of it, afterwards remarking 'the feat could not have been performed without him'.*

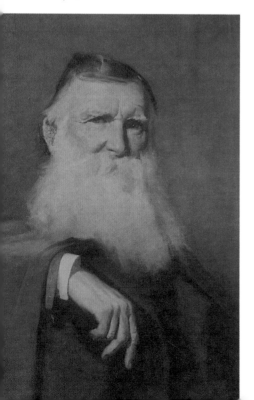

Philip Gidley King, circa 1890s, by an anonymous photographer. King wrote to Darwin from New South Wales (19 April 1863): 'You cannot think with what pleasure I received a note from you by the last Mail enclosing a Photograph of yourself – It was a Strange feeling that came over me as I identified one by one the now well remembered features of what you must permit me to call yr dear old face'.

After retiring from the Royal Navy King stayed in Australia, initially with his father. He married Elizabeth Macarthur in 1843, and had a long and, for the most part, successful career managing a variety of agricultural and business enterprises. In 1851 he was appointed superintendent of stock for the Australian Agricultural Company.

The discovery of gold in 1852 in the Peel River, New South Wales, initially caused panic. King helped to organize, license, and control the gold rush. He became the first mayor of Tamworth, 1876–80, and in 1879 became a commissioner of the Sydney International Exhibition. During the 1880s he was a director of the Mercantile Bank of Sydney.

King wrote *Comments on Cook's Log (HMS Bark Endeavour, 1770) with Extracts, Charts and Sketches* in 1891. He was a life-long friend of Darwin and wrote an unpublished memoir of him in 1892. The manuscript is now in the Mitchell Library, Sydney. He acquired and inherited considerable wealth and left an estate valued at £41,691.

Alexander Burns Usborne (1808–1885) was baptized in Kendal, Cumbria. He joined the navy in 1825 as a second class volunteer and served on HMS *Thetis* until March 1831, latterly as master's assistant. He also served aboard HMS *Revenge* before joining the *Beagle* under FitzRoy. In 1835, with FitzRoy's approval, he took command of the schooner *Constitucion* and surveyed the coast of Peru. As agreed, he did not rejoin the *Beagle* and returned to England in a merchant ship. Known to his friends and shipmates as Jimmy or Gemmy, he was popular. Mount Usborne in Peru was named after him. From 1838 to 1839 he was master of the *Beagle* during her third survey expedition to Australia. He was invalided home after being accidentally shot through the body by a musket ball in Roebuck Bay, Western Australia. He surveyed the coasts of England and Ireland, before he retired with the rank of Captain in September 1867. His recollections and accounts of his service during the *Beagle* expeditions were published in the *Nautical Magazine* in 1838.

Darwin was amused by Midshipman Arthur Mellersh (1812–1894), who truculently introduced himself (according to Darwin family tradition) as: 'I'm Arthur Mellersh of Midhurst [in Sussex], I have read Lord Bryon and I don't care a damn for anyone!'. He entered the navy in 1825 and rose to become an admiral on retirement in 1884. From 28 August 1851 Mellersh was commander of HMS *Rattler*, the experimental British naval vessel powered by screw-propulsion. He participated in the second Anglo-Burmese War (1852–3), and later as captain of HMS *Forte*, the wooden screw-frigate and flagship of Rear-Admiral Richard Laird Warren, he sailed to the south east coast of America. Mellersh died in West Brighton.

FitzRoy's letter of recommendation to the Admiralty for Mellersh was also laudatory:

> ...*he conducted himself with diligence, sobriety and attention and was always obedient to command.... He is a good seaman, an active, careful and extremely correct officer, a good navigator, and had some practice surveying. In his watch a Captain may sleep soundly. No duty that he knows he ought to do is ever neglected by Mr Mellersh.*

Robert McCormick (1800–1890), *Beagle*'s senior surgeon, was without doubt the most unpopular person to be associated with the ship. After McCormick had prematurely left the ship in late April 1832, Darwin described his departure as 'no loss' and a 'political change'.

Darwin had made up his mind before the ship left Devonport that he was not enamoured with McCormick's character. He recalled that all he seemed to care about was 'whether his cabin shall be painted French Grey or a dead white – I hear little excepting this subject from him'. Oddly there is no reference to McCormick's service aboard the *Beagle* in his featured entry within the *Oxford Dictionary of National Biography*.

Richard Darwin Keynes considered him 'an ambitious Scot, determined to make a career for himself'. In fact McCormick was of Anglo-Irish descent and therefore the supposition that he fell out with the ship's captain because

Monday Sept. 26th. Wrote a letter to Capt. Fitz-Roy with a list, suggesting a supply of materials for preserving objects of Natural History. Thursday 29th. Accompanied Capt. Fitz-Roy to the Naval Hospital on a Survey. Visited Dr Armstrong's Collection of Minerals. — Sunday Oct. 2nd. Attended forenoon Service at the Dock Yard Chapel. (Wet day) Friday 7th. Attended a Survey on Warrant Officers at the Commissioner's Office, Naval Hospital. Sunday 9th. Attended Forenoon & afternoon Service at the Dock Yard Chapel. (Wet afternoon.) Thursday 13th. Went to the Naval Hospital in the Boat for the Beagle's — Medicines — blew a hard Gale of Wind with heavy Rain in the afternoon — Saturday 15th — Went to a Concert at the Devonport Town-Hall in the Evening. —

Detail from Robert McCormick's *Rough Diary*. McCormick was the most unpopular crewmember of *Beagle*'s second voyage. But was he unfairly treated? Collecting specimens was traditionally the role of the ship's surgeon

of FitzRoy's Irish ancestry is doubtful. The Wellcome Library in London, 'for the history and understanding of medicine', holds an extensive archive of papers and journals, as well as sketches and drawings by McCormick, including his *Rough Diary of the 2nd Voyage to West Indies 1830 and Voyage to Rio-Janeiro & South America in 1831–2*.

The *Rough Diary* records some fascinating details of the events leading up to the *Beagle*'s departure from Plymouth. McCormick also made many observational records, especially of birds, including a detailed list of those shot. His Diary records the timeframe of his premature exit. On 15 April 1832 he had made his application to be 'invalided'. A week later he attended an 'invaliding survey' on board HMS *Warspite*, and by 23 April he had received the 'invaliding certificate' allowing him to return home.

McCormick stated his reasons for leaving the *Beagle* in his two-volume publication *Voyages of discovery in the Arctic and Antarctic seas, and Round the world*, published in 1884:

> Having found myself in a false position on board a small and very uncomfortable vessel, and very much disappointed in my expectations of carrying out my natural history pursuits, every obstacle having been placed in the way of my getting on shore and making collections, I got permission from the admiral in command of the station here to be superseded and allowed a passage home in H.M.S. Tyne.

Darwin's letter to his Cambridge mentor, Professor John Stevens Henslow, written at Rio de Janeiro, 18 May 1832, gives his reason why McCormick left: 'I am very good friends with all the officers; & as for the Doctor he has gone back to England – as he chose to make himself disagreeable to the Captain & to Wickham. He was a philosopher of rather an antient date'. FitzRoy's letter to Beaufort from Montevideo on August 1832 make his views on McCormick clear: 'We were fortunate in so soon parting company with the surgeon; he was a sad empty headed coxcomb'.

McCormick may well have been a quarrelsome character; however, he cannot be taken to task for being ambitious and clearly his hopes of being recognized as a prominent naturalist and collector of specimens (traditionally part of the senior surgeon's duties onboard a ship) were severely hindered by the energetic and intelligent landsman Charles Darwin who was given every assistance in his work. McCormick resented Darwin's preferential treatment. Darwin was the captain's specially appointed naturalist and companion, and a gentleman of a comparable social status. FitzRoy did not perceive

McCormick as his social equal. The captain was helping Darwin accumulate, and ship back home, a private collection subsidized by the British public purse, although Darwin would ensure that it would be used for public benefit.

Before *Beagle* commenced her survey voyage we know that relations between McCormick and the ship's captain were cordial and they were often in company together. If FitzRoy had his doubts at this early stage he would have prevented McCormick from participating in the voyage. On Saturday, 17 November 1831, McCormick 'ordered a compass and thermometer at Cox's the optician's [and] went to Plymouth Hospital about the *Beagle*'s supply of medicines'. He requested additional medicines from the Victualling Board and wrote a letter to FitzRoy with a 'list suggesting a supply of material for preserving objects of Natural History'.

Earlier, on 13 November, he noted that 'in the evening Capt Fitz Roy brought the Fuegians on board'. He saw Fuegia Basket at Weatherley's Hotel on the following day. On the 15th McCormick finally received his orders for the *Beagle*'s voyage 'to sail between the 30th of Nov and the 3rd of December'. In fact, because of inclement weather the *Beagle* would not actually depart until 27 December.

Robert McCormick, lithograph after Stephen Pearce (1819–1904). The original oil painting by Pearce is now part of the collections of the National Portrait Gallery, London.

McCormick studied surgery at Guy's and St Thomas's hospitals, London, under Sir Astley Cooper (1768–1841). He was the star pupil and gained his diploma in 1822. He became a naval surgeon in 1823. Four years later he joined William Edward Parry's polar expedition to the north of Spitsbergen.

Between 1839 and 1843 McCormick joined the Antarctic expedition in HMS *Erebus* under the command of James Clark Ross (1800–1862), as surgeon and naturalist. Eventually in 1852, after extolling the virtues of a small-boat search expedition to the Arctic, he was given the opportunity to organize a search expedition for Sir John Franklin (1786–1847), who had taken HMS *Erebus* and HMS *Terror* to search for the elusive North-West Passage. The expedition had set off from Greenhithe in Kent on 19 May 1845 and never returned.

McCormick wrote an account of his polar experiences and was awarded the Arctic Medal in 1857. He was constantly canvassing, and was in conflict with, the Admiralty over his career progression and related promotions. He consoled himself with extended walking holidays in which he described his appearance and apparel: 'dressed in a Florentine shooting-jacket and Swiss checked trousers with a change of linen in my pocket, compass and thermometer, road-book and note book'. But he died a

Syms Covington, by an anonymous photographer. Covington was Darwin's shipboard and land assistant. After emigrating to Australia he corresponded with Darwin and sent him specimens to assist his work.

Benjamin Bynoe (1803–1865) served on all three of *Beagle*'s survey voyages in a medical capacity. He deserved then, and today, greater recognition and praise for his contribution to British naval survey work and as a collector of significant specimens too. His collection of 'birds, fish, coleoptera [beetles and insects] and Lepidoptera [butterflies and moths]' found its way into the British Museum (Natural History) in London. However, only one specimen named after him is widely remembered today, the shrub *Acacia Bynoeana*. He faired better in terms of geographic place names and is remembered by Bynoe Inlet, Bynoe Island and Cape Bynoe. A significant part of the collections assembled by Bynoe on the third (and last) of *Beagle*'s survey expeditions was presented to the Royal Botanic Gardens at Kew.

Bynoe was one of the most popular of FitzRoy's crewmembers, who benefited from McCormick's resignation as *Beagle*'s senior surgeon. Darwin thought very highly of Bynoe and in the preface to the third volume of the *Narrative* wrote: 'I must likewise take this opportunity of returning my sincere thanks to Mr. Bynoe, for his very kind attention to me when I was ill at Valparaiso'. Darwin appeared well during his visit to Santiago and rode and climbed with gusto in the central Cordillera of the Andes. On 19 September 1834 he wrote: 'During the day I felt very unwell, and from that time to the end of October did not recover'. Darwin thought that his illness had been brought on by drinking Chi Chi, 'a very weak, new-made, sour wine'. Bynoe treated him with calomel. FitzRoy was greatly concerned and kept the *Beagle* at Valparaiso so that Darwin could be treated ashore to aid his recovery.

Bynoe was born on 25 July 1803 in the parish of Christ Church of Barbados. He left this British colony to study medicine in Britain. He obtained his London Diploma on 18 March 1825 and on 20 May of that year received membership of the Royal College of Surgeons of England. He passed as Assistant Surgeon in the Royal Navy and was given a position as a supernumerary onboard HMS *Victory*. A month later he was instructed to sign on for HMS *Beagle*'s first survey expedition as an assistant to the senior surgeon.

By July 1828 Bynoe was appointed Acting Surgeon, as his predecessor had left through illness. With the suicide of Pringle Stokes, *Beagle*'s captain, the command passed to FitzRoy, who recognized Bynoe's skills and actively encouraged him. He ensured that he was part of his survey team for the second voyage, albeit initially in a secondary role to McCormick. During *Beagle*'s third voyage Bynoe was senior surgeon.

Bynoe had been of great help to the Fuegians during their residence in England and the second voyage. He had become 'their most confidential friend'. FitzRoy openly acknowledged Bynoe's assistance on land and at sea. Bynoe was a good shot and would often accompany the captain during landing parties and act as a scout and bodyguard. FitzRoy acknowledged that his 'affectionate kindness of Mr. Bynoe on…every occasion when his skill and attention were required, will never be forgotten by any of his shipmates'. Bynoe helped the captain as best as he could through his periods of depression and ill health by making him feel comfortable, but in truth, there was nothing he could have done, for his medical condition at this time was not understood and, as such, was untreatable.

After Bynoe's association with the *Beagle* ceased he returned to his home in the Old Kent Road in London and made do on naval half-pay. There was some good news: in 1844 the Royal College of Surgeons elected him as a Fellow. He was finally appointed as Senior Surgeon and commissioned on several convict transports sailing from England to Australia. These long, arduous voyages started to take their toll on his health and he became seriously ill on the *Lord Auckland* which left England bound for Hobart on 19 March 1846.

Charlotte, his loyal and supportive wife, tried to gain more suitable appointments for her husband but to no avail. Bynoe was also employed onboard ships sailing to Cork 'to aid in carrying out measures for the relief of the *Distressed Irish*'. This was during the period of the potato blight and wide-scale famine in Ireland that lasted from 1844 to 1849.

On 23 January 1863 Bynoe, almost 60, was officially recorded as retired by Admiralty Order. For someone who contributed so much to the success of *Beagle*'s voyages Bynoe left a modest estate valued at less than £450.

Syms Covington (1813–1861) also deserves greater acclaim for his contribution on the *Beagle*. He was originally signed onboard the ship by FitzRoy as 'fiddler and boy to the poop cabin'. It transpired that he was partially hard of hearing and there is no record of his musical ability.

Before Darwin sought formal approval from his captain, Covington helped him to record and store his specimens on a casual basis. FitzRoy was clearly aware that Darwin needed assistance, and so Covington was eventually approved as his servant and assistant. Darwin (or rather his father) paid him an annual salary of £60. Writing to thank his father, Darwin was confident that he would 'now make a fine collection in birds and quadrupeds, which before took up far too much time'.

Darwin added a curious comment about Covington in this letter: 'My servant is an odd sort of person: I do not very much like him: but he is perhaps from this oddity, very well adapted to all my purposes'. There is no evidence to explain the oddity but it did not prevent them forming a long-lasting working relationship. Darwin taught Covington to shoot, collect, preserve and pack specimens. As there was a standing order onboard ship forbidding solo land excursions Darwin benefited from Covington's company. It gave him greater independence.

Walking Dress of the Females of Lima
(Peru) August 1835

Covington was a landsman from Bedfordshire when he joined the *Beagle* on her second survey voyage, appearing on the muster book as a 'Boy 2nd Class'. He appeared at various times to have been a very reluctant crewmember who missed home life. Covington wrote a journal of his observations and experiences on the voyage. Remarkably it has survived and provides a complementary account from a very different perspective to those created by Darwin and FitzRoy. He was also an amateur artist and more than twenty examples of his work have survived, including coastal profiles, landscape views, and notably, several portraits of the exotic dress of the ladies of Lima, Peru.

Of Port Desire in Tierra del Fuego Covington wrote:

> *…is much the same as other parts of Patagonia, viz. sandy hills with very bad brackish water, and that obliged to dig for; but some valleys are very pleasant: in season there are plenty of wild cherries. They were nearly ripe at this time.… Birds are not so numerous nor so splendid here as in many other parts of South America, but of course they are less well known. About this part no deer were seen, but immense quantities of guanacos, also lions, foxes, ostriches and aperea or guinea pig. The cliffs are full of fossil shells.*

Covington was fascinated by buildings, farms and land, but he was not impressed with the gaudy and costly interiors of Catholic churches, describing them as 'more like fairy castles than places of Divine worship'. As part of his work he was to provide meat, vegetables and drinking water for Darwin and so not surprisingly many of his journal references relate to these routine labours. In addition, he also displays more than a passing interest in the wildlife and geology of certain areas. Covington may have been involved in collecting specimens, too. He certainly spent a great deal of time and effort packing them for safekeeping onboard *Beagle* and transportation on homebound ships.

After the voyage Covington continued to work for Darwin as his assistant, secretary and servant. In 1840 he emigrated to New South Wales. Darwin had provided a letter reference (now in the Mitchell Library, New South Wales) for him sealed in red wax stating that he 'accompanied me as servant in the voyage of the H.M.S. Beagle, around the world and made himself generally useful. He assisted me as clerk'. Covington later became postmaster of the settlement of Pambula, Twofold Bay, in New South Wales and also managed an inn called the Retreat.

Darwin remained in contact with Covington, who sent him samples of local barnacles and other specimens to assist his research work. Some of them can now be seen in the British Museum. Darwin sent him a new ear trumpet to help his hearing. He died of 'paralysis' in his late forties.

Bearing in mind that so many of the seamen who served with FitzRoy went on to distinguish themselves in various careers it would not be an exaggeration to suggest that was in no small part due to the influence, leadership, and personal encouragement of *Beagle*'s captain.

Opposite: A Lady of Lima, Peru, watercolour by Syms Covington. Darwin was very taken with these ladies and pondered the possibility that ladies back in England should consider such alluring dress.

Chapter 4
Charles Darwin: Naturalist & Companion

Darwin is a very sensible, hard-working man and a very pleasant messmate. I never saw a 'shore-going fellow' come into the ways of a ship so soon and so thoroughly as Darwin. I cannot give a stronger proof of his good sense and disposition than by saying that 'Everyone respects and likes him.' – Captain Robert FitzRoy, in a private letter to Captain Francis Beaufort

Charles Darwin, watercolour by George Richmond, 1840. The artist was part of an artistic group known as The Ancients, which included Edward Calvert (1799–1883) and William Blake (1757–1827), who was also a celebrated poet, printmaker and visionary.

Charles Darwin was selected by Captain Robert FitzRoy as *Beagle*'s naturalist, and as his messmate and companion. Darwin was a landsman of a comparable social class to FitzRoy, and one in whom he could confide without breaking British naval protocol.

The remarks and information contained in letters and accounts by the key participants of the *Beagle* expedition reveal vivid insights into FitzRoy's personality and behaviour. He was a man who suffered extreme mood swings and was feared by some for his hot temper. The overwhelming evidence, during and after the voyage, points to a man who suffered from a medical condition akin to clinical depression. It is an extraordinary coincidence that Pringle Stokes, *Beagle*'s first captain, and Owen Stanley, the captain of one of her sister-ships, also suffered from mental disorders and depression.

FitzRoy had taken over command from *Beagle*'s first captain during her first expedition. He occupied Captain Pringle Stokes' cabin, the confined space where for a fortnight Stokes had incarcerated himself, and then placed a pistol to his head and pulled the trigger. As a self-aware, highly sensitive and intelligent man, FitzRoy would surely have seen in Darwin someone who could admirably fulfill a dual role: naturalist and gentleman-companion. But to date no written evidence has come to light to confirm that these thoughts were actually on the captain's mind.

For a professional naval officer still in the prime of life it is hardly surprising that FitzRoy makes no reference in the *Narrative* to his depressive nature and Darwin's dual role onboard. He focuses solely on Darwin's position as a naturalist, writing:

Anxious that no opportunity of collecting useful information, during the voyage, should be lost; I proposed to the Hydrographer

[Beaufort] that some well-educated and scientific person should be sought for who would willingly share such accommodations as I had to offer, in order to profit by the opportunity of visiting distant countries yet little known. Captain Beaufort approved of the suggestion, and wrote to Professor Peacock, of Cambridge [professor of mathematics], who consulted with a friend, Professor Henslow, and he named Mr. Charles Darwin, grandson of Dr. Darwin the poet, as a young man of promising ability, extremely fond of geology, and indeed all branches of natural history. In consequence an offer was made to Mr. Darwin to be my guest on board, which he accepted conditionally; permission was obtained for his embarkation, and an order given by the Admiralty that he should be borne on the ship's books for provisions. The conditions asked by Mr. Darwin were, that he should be at liberty to leave the Beagle and retire from the Expedition when he thought proper, and that he should pay a fair share of the expenses of my table.

In March 1840 the celebrated portraitist George Richmond (1809–1896) was commissioned by Darwin's uncle to paint watercolour portraits of Charles and his wife Emma shortly after their marriage. They are still hanging in the living room of their family home at Down House in Kent. Darwin, aged 31, has an open, fresh face, and although sideburns are evident, he is notably lacking the later trademark full beard. The firmly fixed and animated eyes reveal a lively and enquiring mind. He would benefit from the *Beagle* voyage, reaping fame and fortune. His theories on natural selection – what became known as the 'survival of the fittest' (a phrase coined by the British economist Herbert Spencer (1820–1903) who was inspired by Darwin) – derived directly from his observations and collections gathered during the expedition. Darwin would become one of the world's most famous men of science.

FitzRoy was a firm believer in the pseudo-science of phrenology, invented by the German physician Franz Joseph Gall (1758–1828) around 1800. Gall originally called it cranioscopy and it was later renamed phrenology: FitzRoy called it 'bumpology'. Phrenology advocated that it was possible to determine someone's character and personality from the shape of their head by reading the 'bumps' and 'fissures'. FitzRoy was initially wary of the shape of Darwin's nose but eventually decided he would be suitable for the shipboard position.

On the face of it Darwin seemed an unlikely choice for such an appointment as, by his own admission, he had squandered his time at Cambridge, where his interests were predominantly in hunting and shooting. He went up to Christ's College for the purpose of becoming a clergyman. Earlier he had failed to qualify as a doctor because he was unable to stomach the anatomy lessons and operations at Edinburgh University.

Dr Robert Waring Darwin, by Ellen Sharples. The doctor was so large, a servant had to enter a patient's home before him and jump up and down on the floorboards to ensure they would withstand his weight.

Darwin's father, Robert, was physically the largest man Darwin had ever seen. He was about 6 feet 2 inches in height, and well before middle age he tipped the scales at 24 stones. The weight gain continued apace. He was a highly respected, kindly, intuitive and very successful society doctor and financier, with significant interests in farm estates, railways and canals, who was very concerned about his sons' future prospects. In fact Charles, and his siblings were remarkably fortunate in that theoretically they never had to work for a living.

Charles recorded his father's anxieties and concerns in his *Autobiography* (begun in 1876 and published posthumously in 1887):

> *When I left school I was for my age neither high nor low in it; and I believe that I was considered by all my masters and by father a very ordinary boy, rather below the common standard in intellect. To my deep mortification my father once said to me, 'You care for nothing but shooting, dogs, and rat-catching, and you will be a disgrace to yourself and all your family'.*

Charles Robert Darwin was born on 12 February 1809 at The Mount, a large Georgian house overlooking a bend in the River Severn in Shrewsbury. He was the second son and fifth (of six) children to Dr Robert Waring Darwin (1766–1848) and Susannah Darwin, née Wedgwood (1765–1817). He was the grandson of Dr Erasmus Darwin (1731–1802), a larger-than-life character in every respect, who FitzRoy called 'Dr Darwin the poet', and of Josiah Wedgwood I (1730–1795), who established the famous pottery factory at Etruria near Stoke-on-Trent. The factory is credited with the industrialization of the production of pottery. They were renowned gentleman scientists, inventors and original thinkers. Poetry was only a small part of Erasmus's many and varied interests.

Darwin's grandfathers were both Fellows of the Royal Society and founding members of the Lunar Society of Birmingham (so-called because they met on nights when the full moon provided light for the journey home). In the late 1700s a small group of 'Lunaticks' assembled for what Erasmus called 'a little philosophical laughing'. They included Matthew Boulton (1728–1809), the engineer and manufacturer, Joseph Priestley (1733–1804), the chemist who isolated oxygen, and James Watt (1736–1819), the Scottish inventor of the steam engine.

Erasmus's book *Zoonomia or the Laws of Organic Life*, published in 1794–6, pre-dated by several years the ground-breaking publication by the French scholar Jean-Baptist Lamarck (1744–1829), Professor of Zoology at the National History Museum in Paris, in rejecting the doctrine of creationism. Erasmus was one of the first scientists to argue that organisms and species could be gradually transformed through the influence of their needs. This concept would have a

Opposite top: The Wedgwood Family, oil painting dated 1780, by George Stubbs (1724–1806). Josiah Wedgwood I and his wife are portrayed to the far right. Darwin's mother, Susannah, is on the horse in the centre, and next to her looking out is the young man Josiah II, or 'Uncle Jos', the father of Darwin's wife Emma. Wedgwood liked the horses but not the portraits of his family, which he regarded as 'wooden mannequins'

Below: Dr Erasmus Darwin, oil after Joseph Wright of Derby. Derby captured on canvas many of the people, and experiments, involved in the early phases of the Industrial Revolution.

significant influence on Darwin's theories. Erasmus was painted by his friend, Joseph Wright of Derby (1734–1797), who was also one of his patients. Many of Wright's celebrated pictures related to experiments performed by members of the Lunar Society.

Josiah Wedgwood's eldest child, Susannah, had married Darwin's father. Susannah's brother, Josiah Wedgwood II (1769–1843), was affectionately known in the Darwin household as Uncle Jos. He was the best friend of Charles's father and he took over the management of the pottery business. But Uncle Jos lacked his father's business acumen and he was fortunate to have Robert Darwin to provide assistance and financial help. Robert lent him £30,000 for the purchase of the Elizabethan mansion of Maer Hall in Staffordshire. Charles was a regular visitor, where he indulged his passion for hunting and shooting. There was a lake and about a thousand acres of land. For Charles it was 'Bliss Castle'.

Another place Charles enjoyed visiting was Woodhouse, a short ride north of Shrewsbury, and the home of the Owen family. William Owen was a friend of his father and a business client, too. He was fond of Charles, and the boy loved the informality of the household. William encouraged him to develop his marksmanship skills, and one of his daughters, Fanny, a lively, pretty girl, caught Charles's eye. They corresponded amicably but before he returned from the *Beagle* voyage Charles was dismayed to learn that she had married.

In July 1817 his mother died. Charles was too young to remember much about her. He was sent to board at Shrewsbury School, which he hated, and

Above: Charles Darwin and Catty by Ellen Sharples. Painted before his mother's death, Darwin is believed to be around 7 years old. He holds a plant alongside his youngest sister.

where he failed to shine academically. Fortunately it was only a short run from his home and he returned as often as he could. At that time the curriculum focused on the classics, and science was not taught there; however Charles, and his older brother Erasmus (1804–1881), set up a 'Laboratory' in an outhouse of their family home where they performed makeshift experiments examining the properties and compositions of various domestic items. To do this, Charles often subjected the material to a naked flame, which earned him the childhood nickname 'Gas'. Erasmus went up to Edinburgh to study medicine but his letters home to Charles encouraged him to continue with the experiments.

By all account at this stage of his life there is no evidence of the genial Charles (or as his father and family called him, Charley or Bobby) that would later emerge. He was an inward-looking, slightly stammering child, like his father and grandfather before him, who preferred to live in a semi-fantasy world. At University his personality radically changed and he became more extrovert in nature.

A charming double portrait by the (British-born) artist Ellen Sharples (1769–1849), who was active in America, shows Charles with his youngest sister, Emily Catherine (Catty), clutching a plant, painted the year before their mother's premature death. The Mount boasted an impressive collection of rare floral specimens. Darwin's older sisters, Marianne, Caroline and Susan, spoiled and mothered him. That he was close to all of them is evident from the large number of detailed and affectionate letters sent home from the *Beagle*. His father was a great talker, who liked to dominate conversations. But he was not a letter writer and so his messages were conveyed via his daughters' epistles.

Robert Darwin was determined that Charles should study medicine and sent him to Edinburgh University, where he was to accompany Erasmus who was

View of Christ's College, Cambridge, by John Le Keux from *Memorials of Cambridge* (1837). College tradition asserts that Darwin's rooms at Cambridge were earlier occupied by the seventeenth-century poet John Milton (1608–1674), celebrated for his epic poem 'Paradise Lost'.

pursuing the same goal. In fact, Charles was named after his father's elder brother, who had also studied medicine at Edinburgh but died of an infection caught through performing an autopsy. Erasmus graduated, although he later abandoned the profession, much to his father's displeasure, and never secured gainful employment. Charles had no interest in becoming a doctor. He, among many others, was bored by the lessons of the Scottish anatomy teacher, Alexander Monro. Robert himself had been forced into the profession by his own father. He detested the sight of blood but he continued as a doctor, although the money was but a small proportion of his substantial annual income derived from various financial and property investments.

Charles's time at Edinburgh was not completely wasted. He and Erasmus enjoyed the chemistry lectures and demonstrations given by Professor Thomas Charles Hope (1766–1844). Charles was taught taxidermy by John Edmonstone, the freed black slave from Guyana, South America, who in turn had learned his craft from the naturalist and explorer Charles Waterton (1782–1865). Edmonstone fired up Darwin's imagination with images of the tropical rainforests of his homeland. He attended Robert Jameson's (1774–1854) extra-curricular lectures, which included botany, geology, hydrography, meteorology, mineralogy and zoology. Charles's personal, annotated copy of Jameson's *Manual of Mineralogy* (1821) became an important influence on his revolutionary theories. But he was not impressed with the Scottish Professor, describing him as 'that old brown dry stick', although the survival of his undergraduate notes reveals that the lessons were, according to Janet Browne, the acclaimed Darwin biographer, 'both comprehensive and useful'.

Charles was an active member of the Plinian Society where he met Dr Robert Grant, a lecturer on invertebrate animals. Many years later he recalled that:

> *He one day, when we were walking together, burst forth in high admiration of Lamarck and his views on evolution. I listened in silent astonishment, and as far as I can judge, without any effect on my mind. I had previously read the* Zoonomia *of my grandfather, in which similar views are maintained, but without producing any effect on me. Nevertheless it is probable that the hearing rather early in life such views maintained and praised may have favoured my upholding them under a different form in my* Origin of Species. *At this time I admired greatly the* Zoonomia; *but on reading it a second time after an interval of ten or fifteen years, I was much disappointed, the proportion of speculation being so large to the facts given.*

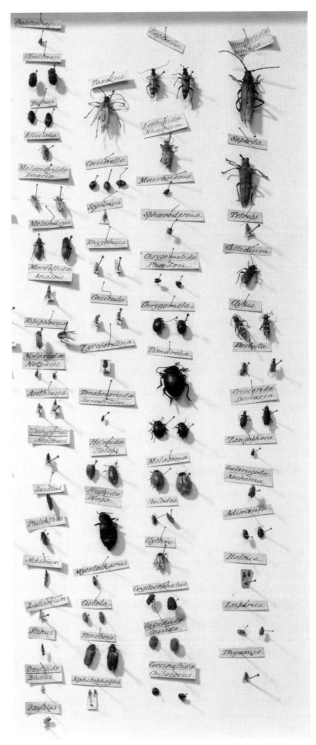

A case of beetles from Down House. Darwin was passionate about collecting beetles. Albert Way, a fellow Cambridge student, portrayed Darwin in two light-hearted pen-and-ink caricatures riding atop a beetle. They are now in the Cambridge University Library.

Reverend John Stevens Henslow, oil painting by an anonymous nineteenth-century artist. Darwin was Henslow's favourite pupil, and was the man responsible for encouraging Darwin to take up the position of naturalist and gentlemen-companion aboard the *Beagle*. The portrait is possibly by Thomas Herbert Maguire (1821–1895), the artist who produced many lithographs after his own portraits. He painted a series of 'Ipswich Museum Portraits' in 1851. The Ipswich Museum was established in 1847 as a broad-based natural history collection. Its origins and development owed much to Henslow, who was elected President in 1850.

Darwin had neither the aptitude nor the inclination to pursue a medical career. He abandoned his studies at Edinburgh and, after what must have been difficult discussions with his father, it was mutually agreed that he would pursue the career choice of a country clergyman. In January 1828 Charles went up to Christ's College, Cambridge, to study for a Bachelor of Arts degree, the entry level to study Holy Orders. In contrast to his childhood personality Darwin by this time was an amiable and affable student and had a remarkable capacity to develop and maintain friends. No doubt Charles would have been popular with his parishioners and concurrently could have pursued scientific interests. However, had he pursued this vocational path he would almost certainly not be the household name of today.

At Cambridge Darwin continued his interests in hunting and shooting. He was also passionate about beetles: 'No pursuit at Cambridge was followed with nearly so much eagerness or gave me so much pleasure as collecting beetles. It was the mere passion for collecting, for I did not dissect them and rarely compared their characters with published descriptions, but got them named anyhow'.

Darwin enjoyed the companionship of his cousin William Darwin Fox (1805–1880), who was also studying at Cambridge and who shared Charles's passion for insects. But without doubt the most influential figure within the University was the Reverend John Stevens Henslow (1796–1861), Professor of Botany. It was Henslow who recommended Darwin for the *Beagle*'s second voyage. Darwin wrote in his *Autobiography* of his friendship with Henslow:

> *During the latter half of my time at Cambridge [I] took long walks with him on most days, so that I was called by some of the dons 'the man who walks with Henslow'; and in the evening I was very often asked to join his family dinner. His knowledge was great in botany, entomology, chemistry, mineralogy, and geology…. His judgement was excellent, and his whole mind well-balanced; but I do not suppose that anyone would say that he possessed much original genius.*

During the early years of the nineteenth century the study of Natural History was still in its relative infancy. Although Darwin had no formal qualifications in these spheres, his knowledge was significant. Henslow believed that he possessed the right qualities to benefit from and significantly contribute to scientific research. As Richard Darwin Keynes aptly surmised, Darwin had 'greatly impressed some of the most eminent scientists in Cambridge with his practical ability as a collector, and with the high quality and purposefulness of his enquiring mind. "What a fellow that Darwin is for asking questions," said Henslow'.

Henslow also encouraged Darwin's interest in geology and Adam Sedgwick (1785–1873), Professor of Geology at Cambridge, became another prominent figure in Darwin's educational development. Darwin had just returned from a field excursion with Sedgwick in Wales, in preparation for a proposed Tenerife expedition, when the news of FitzRoy's search for a 'naturalist' broke.

Darwin, like FitzRoy, was a keen reader. They had both read the travel writings of Alexander von Humboldt (1769–1859). For Darwin, Humboldt's vivid accounts of faraway places fired up his imagination, inspiring him to make plans for an expedition with class mates to explore the island of Tenerife and further afield. Humboldt was a German (Prussian) aristocrat and a remarkably gifted naturalist and explorer, who between 1799 and 1804 had travelled to Latin America and was the first to explore and describe it from a scientific viewpoint. Writing in his Diary Darwin notes: 'At present I talk, & think & dream of a scheme I have almost hatched of going to the Canary Islands…. I have long had a wish of seeing Tropical scenery & vegetation: & according to Humboldt, Teneriffe [one of the Canary Islands] is a very pretty specimen'.

FitzRoy's request gave Darwin the opportunity to turn some of his dreams into reality, although it forced him to abandon his plans of an independent expedition to Tenerife. Henslow's parting gift to Darwin before he departed on the *Beagle* was a copy of an English translation of the first two volumes of Humboldt's *Personal Narrative of Travels to the Equinoctial Regions of the New Continent*. It was inscribed, 'J. S. Henslow to his friend C. Darwin on his departure from England upon a voyage round the world. 21 Sept. 1831'. Darwin made many references to Humboldt in his letters, journals and accounts. His first letter to Henslow from Rio de Janeiro, 18 May 1832, was full of youthful enthusiasm: 'At Santa Cruz, whilst looking amongst the clouds for the Peak & repeating to myself Humbold's sublime descriptions'.

But before Darwin could accept the shipboard position he needed the approval of his father. Initially Robert thought that such a venture would be a worthless and highly dangerous enterprise and could be of no benefit to his son. Fortunately he was persuaded by Uncle Jos, who in letter form had weighed up the pros and cons of the venture. Uncle Jos believed that the 'pursuit of natural history, although certainly not professional, [was] very suitable for a clergyman'.

Darwin, following his father's wishes, had initially written to turn down the position. There was a rival candidate, Leonard Jenyns, who was a clerical naturalist and vicar of Swaffham Bulbeck near Cambridge. He was Henslow's brother-in-law. Jenyns had considered the post carefully and decided that he preferred to remain within his parish. Henslow himself had also pondered the position but acknowledged that he was probably too old, being in his mid-thirties and with responsibilities to his wife and newborn child. So Darwin was back in the running and, with his father now offering full moral and financial support, he eventually secured the position.

Henslow wrote to Darwin on 24 August 1831:
> …*I have stated that I consider you to be the best qualified person I know of who is likely to undertake such a situation. I state this not on the supposition of yr being a finished Naturalist, but as amply qualified for collecting, observing, & noting anything worthy*

Alexander von Humboldt, oil by Friedrich Georg Weitsch (1758–1828) painted in 1806. Darwin idolized Humboldt as a student. His multi-volume work *Kosmos* (unfinished at his death) helped to popularize science.

Top: Darwin's combined compass/clinometer now on display at Down House. The maker is not known but Darwin was given considerable assistance by his Cambridge tutors, family, and by FitzRoy, to equip himself for the *Beagle* expedition.

Bottom: Darwin's geological hammer can also now be seen at Down House. He had been hammering rocks in Wales with Adam Sedgwick, the Cambridge Professor of Geology, when he heard of the offer to join the *Beagle* expedition.

to be noted in Natural History…Capt F [FitzRoy] wants a man (I understand) more as a companion than a mere collector & would not take any one however good a Naturalist who was not recommended to him likewise as a gentleman.

Interestingly in the same letter Henslow believed that the voyage would last only two years and advised Darwin to take plenty of books with him. The voyage would last the best part of five years.

Henslow encouraged Darwin to prepare and equip himself for the expedition advising on the purchase of various scientific instruments. He also demonstrated some of them to Darwin. On his recommendation Darwin had already acquired a compass and clinometer combined within one case for the proposed Tenerife trip. The clinometer was used to measure the angle of a slope, and the inclination of rock beds. It is now on display at Down House. FitzRoy was of the mistaken understanding that Darwin favoured geology and his gift to him on *Beagle*'s departure from Plymouth was a copy of Charles Lyell's *Principles of Geology*, Volume I (1830). Darwin acquired the second and third volumes during the voyage. The reason why Darwin's selection to the *Beagle* post turned out to be such an inspired decision was because he was not a specialist in any particular field. He had an open and enquiring mind and was interested 'in all branches of natural history'. Darwin could envisage the bigger picture.

Darwin later became aware that he was almost not appointed because of his nose:

On becoming very intimate with FitzRoy I heard that I had run a very narrow risk of being rejected, on account of the shape of my nose! He was an ardent pupil of Lavater [Johann Kaspar Lavater, a Swiss theologian who practised phrenology], and was convinced that he could judge a man's character by the outline of his features; and he doubted whether anyone with my nose could possess sufficient energy and determination for the voyage. But I think it was afterwards well satisfied that my nose had spoken falsely.

FitzRoy may have had doubts about taking onboard a complete stranger recommended by the Cambridge men and possibly invented a 'Mr Chester' as a get-out clause, who he claimed was a potential candidate for the post. But after meeting Darwin and clearly getting along well with him, 'Mr Chester' was no longer available for the position.

On the expedition Darwin became known to everyone as the 'philosopher', or 'philos' for short. This was an appropriate solution to a curious situation. Darwin held no rank, but he had the privilege of being able to address the captain by his surname, and it would have seemed odd for the officers and crew to call him Sir, or Mr Darwin. He was a young independent traveller with freedom to come and go, within reason, as he pleased. His father's annual payments to the Admiralty guaranteed this privileged position and also ensured that the specimens collected were assigned as his personal property.

Darwin had an abundance of energy and applied himself in earnest to the extensive preparations before departure. His sisters proved adept at assembling appropriate clothes. In addition to measuring and surveying instruments, he acquired microscopes, including a portable dissecting one by Robert Bancks, who was instrument maker to George IV. Darwin had taken advice before buying from Robert Brown (1773–1858), the pioneering Scottish botanist and microscopist who in 1801 had sailed as the naturalist on Matthew Flinders' (1774–1814) survey voyage to northern and southern Australia in the *Investigator*. Bancks' instruments had served him well. Later in Devonport Darwin heeded to FitzRoy's recommendation and purchased a set of aneroid barometers.

Darwin also bought traps, nets and hooked fishing lines, as well as books and items to store his specimens, which included preserving papers. He received assistance from William Yarrell (1784–1856), the crack shot, accomplished angler, acclaimed naturalist, and astute bargainer, who recommended a rifle and a brace of pistols for what he considered to be the knock-down price of £50, which Darwin thought might come in useful to fight those 'd— Cannibals'. The cost of equipping him for the voyage was about £600. Henslow also provided introductions to various specialists, including veterans of naval voyages such as the Scottish surgeon and polar explorer Sir John Richardson (1787–1865). He recommended that Darwin take a crash course on the rudiments of navigation to understand latitude and longitude.

Simple microscope by Robert Bancks, owned by Darwin and now on display in his study at Down House. Bancks was part of the firm Bancks and Son – Instrument makers and Opticians to His Majesty, based at 119 New Bond Street, London (1820–27).

As a supernumerary on FitzRoy's ship Darwin's position was unpaid and he was expected to contribute £50 per annum to the Admiralty towards his upkeep. His assistant, Syms Covington, received £60 per annum plus expenses. The number of diplomatic references in his letters to his father and sisters relating to money issues and banker's drafts is telling. Writing to Caroline on 9 August 1834, he reassures with the following words:

> *Give my Father my best love & thanks for all his kindness about money, & tell him I can seriously say that since leaving England I have spent none excepting in the furtherance of Natural History, & as little as I could in that, so that my time should not be thrown away.*

Darwin's father spent more than £1,500 on his son's expedition.

By early October 1831 Darwin had said his farewells to family and friends. He received news that the *Beagle* would be delayed in sailing from Devonport, Plymouth, and so he decided to stay in London for a few weeks. On 24 October he finally arrived in Devonport, although inclement weather would prevent the ship from leaving for more than two months. On the 25th, Darwin noted in his Diary:

> *Went on board the Beagle, found her moored to the Active [a former naval ship] hulk & in a state of bustle and confusion.— The men were chiefly employed in painting the fore part & fitting up the*

Cabins.– The last time I saw her [12 October]…she was in the Dock yard & without her masts or bulkheads & looked more like a wreck than a vessel commissioned to go round the world.

Darwin's instruments and books were taken on board and after two days' hard work with his cabin companion John Lort Stokes and brother Erasmus they managed to arrange everything into a 'very neat order'. There was no room to swing the proverbial cat in this living, sleeping and storage space measuring about 10 by 11 feet that Darwin would have been unable to stand upright in.

Darwin's extensive correspondence includes only a brief mention of his living quarters. In a letter to Henslow he observed that, 'The corner of the cabin, which is my private property, is most woefully small. – I have just room to turn round & that is all'. The poop cabin contained not only the drafting table but it was also the repository for the ship's privately owned but collectively enjoyed library that included more than one hundred volumes, such as bibles, dictionaries, encyclopaedias, novels, pamphlets, poetry and reference works. Several reference books helped Darwin to identify fossils and other specimens, although the authors were not always right in their descriptions and details. In addition, FitzRoy had his own library that his companion could consult.

Evidence of FitzRoy's early relationship with Darwin is found in this entry in Darwin's Diary for 22 November, while still at Devonport: 'Went on board & returned in a panic on the old subject want of room, returned to the vessel with Cap FitzRoy, who is such an effectual & goodnatured contriver that the very drawers enlarge on his appearance & all difficulties smooth away'.

Around a fortnight before departure Darwin candidly expressed his concerns about the responsibilities and challenges of his appointment:

An idle day; dined for the first time in Captain's cabin & felt quite at home. Of all the luxuries the Captain has given me, none will be so essential as that of having my meals with him. I am often afraid I shall be quite overwhelmed with the numbers of subjects which I ought to take into hand. It is difficult to mark out any plan & without method on ship-board I am sure little will be done. The principal objects are 1st, collecting, observing & reading in all branches of Natural history that I possibly can manage. Observations in Meteorology. French & Spanish, Mathematics & a little Classics, perhaps not more than Greek Testament on Sundays. I hope generally to have some one English book to hand for my amusement, exclusive of the above mentioned branches. If I have not energy enough to make myself steadily industrious during the voyage, how great & uncommon an opportunity of improving myself shall I throw away. May this never for one moment escape my mind, & then perhaps I may have the same opportunity of drilling my mind that I threw away whilst at Cambridge.

FitzRoy's disciplined naval training and shipboard routine no doubt rubbed off on his messmate. Admiralty Regulations of this period insisted that a ship's captain kept a daily log with detailed notes and accounts of important incidents and events. Darwin may well have had his own ideas about keeping written records, but certainly FitzRoy's example reinforced Darwin's notions about maintaining regular records of his observations. Darwin's excessive workload of collecting, describing, preserving and packing specimens was formally acknowledged by FitzRoy and he obtained official approval for Darwin's request for an assistant. Syms Covington was formally appointed to that position and became an indispensable aid to Darwin.

On 6 January 1832 the *Beagle* arrived at the Canary Islands and anchored off the port of Santa Cruz on Tenerife. This was Darwin's dream destination but neither he nor any of the seamen went ashore. Reports of cholera in England worried the port officials and the British Vice-Consul conveyed the news that they could only land if they were first subjected to twelve days of quarantine afloat. Everyone was disappointed but the decision was made to sail on to the Cape Verde Islands arriving at St Jago (Santiago) on 16 January 1832.

Darwin's letter to his father, written on 10 February 1832 two days sail from St Jago, is revealing:

> *The voyage from Tenerife to St Jago was very pleasant & our three weeks at it have been quite delightful. St Jago although generally reckoned very uninteresting was the most exciting. Of course the little Vegetation that there was, was purely tropical. And my eyes have already feasted on the exquisite form & colours of Cocoa nuts, Bananas & the beautiful orange trees. Hot houses give no idea of these forms, especially orange trees, which in their appearance are as widely different & superior to the English ones as their fresh fruit is to the imported. Natural History goes on excellently & I am incessantly occupied by new & most interesting animals. There is only one sorrowful drawback, the enormous period of time before I shall be back in England. I am often quite frightened when I look forward. As yet everything has answered brilliantly, I like everybody about the ship, & many of them very much. The Captain is as kind as he can be. Wickham is a glorious fine fellow and what may appear quite paradoxical to you is that I literally find a ship (when I am not sick) nearly as comfortable as a house. It is an excellent place for working & reading, & already I look forward to going to sea, as a place of rest, in short my home.*

Prophetically he added the following words:

> *I am thoroughly convinced that such a good opportunity of seeing the world might not [come] again for a century. I think, if I can be so soon judge, I shall be able to do some original work in the Natural History — I find there is little known about many of the tropical animals.*

In mid-February Darwin was impressed with the sight of the St Paul's Rocks close to the line of the equator but the crossing-the-line ceremony, the 'watery ordeal', was not to his liking. On the 17th: 'We have crossed the Equator, & I have undergone the disagreeable operation of being shaved'. He went on:

Before coming up, the constable blindfolded me & thus lead along, buckets of water were thundered all around; I was then placed on a plank, which could be easily tilted up into a large bath of water. They then lathered my face & mouth with pitch & paint, & scraped some of it off with a piece of roughened iron hoop: a signal being given I was tilted head over heels into the water, where two men received me & ducked me.

Writing to his father in a letter started on 8 February 1832 and completed later in that month Darwin notes:

The time has flown away most delightfully, indeed nothing can be pleasanter; exceedingly busy, & that business both a duty & a great delight. I do not believe I have spent one half hour idly since leaving Tenerife: St. Jago has afforded me an exceedingly rich harvest in several kinds of Nat.History…. Whenever I enjoy anything I always either look forward to writing it down, either in my log book (which increases in bulk) or in a letter.

In the same letter Darwin makes some insightful comments about the equipment aboard the *Beagle*, his developing collections, and the relationship of the limited space within the ship to his working methods:

I find my collections are increasing wonderfully, & from Rio I think I shall be obliged to send a Cargo home. All the endless delays which we experienced at Plymouth have been most fortunate, as I verily believe no person ever went out better provided for collecting & observing in the different branches of Natural History. In a multitude of counsellors I certainly found good. I find to my great surprise that a ship is singularly comfortable for all sorts of work. Everything is close at hand, & being cramped makes one so methodical, in the end I have been a gainer.

Darwin was a prolific letter writer. This lengthy letter also included the warning to his father that he must 'excuse these queer letters & recollect they are generally written in the evening after my days work. I take more pains over my Log Book, so that eventually you will have a good account of all the places I visit'. His family had knowledge of *Beagle*'s voyage itinerary. Darwin's letters home forewarned his family of the ship's progress and destinations, and the British broadsheet newspapers also featured shipping news. Writing to Darwin his family sent the letters ahead to various British Admiralty stations and remarkably most of the letters arrived intact courtesy of naval ships.

On 28 February *Beagle* was close to the coast of Brazil on the northern side of the 'antient town of Bahia or San Salvador'. Darwin compared the unreality of the scenery to the dramatic and sensational biblical paintings of the artist John 'Mad' Martin (1789–1854). He was of the opinion that you had to see the view to believe it. To an extent he found his favourite travel writer's observations reassuring: 'from what I have seen Humboldt's glorious descriptions are & will for ever be unparalleled: but even with his dark blue skies & the rare union of poetry with science which he so strongly displays when writing on tropical scenery, with all this falls far short of the truth.'

Opposite: 'Corcovado Mountain, Rio de Janeiro', engraving after Augustus Earle from the *Narrative* (1839). Earle was fascinated by the mountain and he accompanied Darwin on a climb to the summit. Atop the mountain today is a statue of Christ the Redeemer that has become an icon of Rio and Brazil.

Darwin was very impressed with Brazil and when they arrived at Rio de Janeiro he made detailed descriptions and stayed ashore for some time with FitzRoy's artist, Augustus Earle, who acted as a guide as he had been resident there on earlier, independent explorations. The scenery of Brazil was exceptional but Darwin found dealing with the people a frustrating experience. On 6 April 1832 he had wasted an inordinate amount of time trying to gain the passports for his expedition into the country's interior. He wrote:

> It is never very pleasant to submit to the insolence of men in office; but to the Brazilians, who are as contemptible in their minds as their persons are miserable, it is nearly intolerable. But the prospect of wild forests tenanted by beautiful birds, Monkeys & Sloths, & lakes by Cavies & Alligators, will make any Naturalist lick the dust even from the foot of a Brazilian.

Darwin made the most of every opportunity to explore and was not reticent about participating in land or sea exploration and survey parties even if it put his life in peril.

Writing to Caroline he thought that they should

> …stay more than a month at Rio. I have some thoughts, if I can find tolerably cheap lodgings, of living in a beautiful village about 4 miles from the town. It would be excellent for my collection & for knowing the Tropics, moreover I shall escape the cauking [caulking] & painting & various bedevilments which Wickham is planning. The part of my life as sailor (& I am becoming one i.e. knowing the ropes & how to put the ship about &c) is unexpectedly pleasant.

He had a few more surprises in store for Caroline, telling her on 8 April 1832: 'You will be terrified at the thought of my combating with alligators & Jaguars in the wilds of the Brazils'.

In his Diary in early May Darwin noted:

> These days have been gliding away; there have been torrents of rain, & the fields are quite soaked with water; if I had wished to walk it would have been very disagreeable, but as it is, I find one hour's collecting keeps me in full employment for the rest of the day. The naturalist in England, in his walks, enjoys a great advantage over others in frequently meeting with something worthy of attention; here he suffers a pleasant nuisance in not being able to walk a hundred yards without being fairly tied to the spot by some new & wondrous creature.

From an effusive letter to Caroline from Botofogo Bay on 25 April 1832, Darwin confides that the captain

> …has communicated to me an important piece of news: the Beagle on the 7th May sails back to Bahia. The reason is a most unexpected difference is found in the Longitudes. It is a thing of great importance & the Captain has written to the Admiralty accordingly. Most likely I shall live quietly here, it will cost a little but I am quite delighted at

the thought of enjoying a little more of the Tropics. I am sorry the first part of this letter has already been sent to the Tyne [HMS Tyne]; I must tell you for your instruction that the Captain says, Miss Austens [Jane Austen] novels are on everybody['s] table…

By 9 May 1832 Darwin noted in his Diary that there were now four *Beagle* men living on shore while *Beagle* continued her survey work. The Fuegians had not yet been settled in Tierra del Fuego, and Darwin revealed a keen sense of humour at the expense of one of the them: 'Earl, who is unwell & suffers agonies from the Rheumatism. The serjeant of Marines, who is recovering from a long illness, & Miss Fuegia Basket, who daily increases in every direction except in height'.

Darwin kept up his correspondence with Henslow from Rio de Janeiro:
I thought of the many most happy hours I have spent with you in Cambridge. I am now living at Botofogo, a village about a league from the city, & shall be able to remain a month longer. The Beagle has gone back to Bahia.… Our Chronometers at least 16 of them, are going superbly; none on record ever have gone at all like them.

But collecting, packing and shipping specimens was far from easy:
I have determined not to send a box till we arrive at Monte Video – it is too great a loss of time both for Carpenters & myself to pack up whilst in harbor. I am afraid when I do send it, you will be disappointed, not having skins of birds & but very few plants, & geological specimens small: the rest of the things in bulk make very little show.

Darwin relayed to Catty (his youngest sister) that on 6 July they would touch at Cape Frio, where HMS *Thetis*, with a substantial amount of money and treasure, had foundered in 1830. He noted:
They have fished up 900,000 dollars. If we are lucky enough (& it is very probably) to have a gale off St. Catherine's we shall run in there. I expect to suffer terribly from sea-sickness – as we are certain to have bad weather. After lying a short time at MV [Montevideo]: we cruize to the South – but not I believe Rio Negro. The geography of this country is as little known as the interior of Africa. I long to put my foot, where man had never trod before – and am most impatient to leave civilized ports.

In fact they would call at Rio Negro. Darwin was involved in quelling a rebellion at Montevideo. In 1821 the Brazilian Emperor Dom Pedro I had annexed the eastern part of Uruguay. Juan Antonio Lavalleja led the insurgents and fought back. Finally in 1828 Uruguay's independence was recognized. The fighting that Darwin witnessed was between rival political parties.

Writing in a letter of 31 July 1832:
We all thought we should at last be able to spend a quiet week, but alas the very morning after anchoring a serious mutiny in some black troops endangered the safety of the town. We immediately arrived & manned all our boats, & at the request of the inhabitants occupied the

Overleaf: 'Botofogo Bay, Rio de Janeiro', watercolour by Conrad Martens. Painted before Martens formally joined the *Beagle* expedition as Augustus Earle's replacement. Darwin shared a cottage with Earle at Botofogo Bay in 1832.

'Monte Video – Custom House', engraving
after Augustus Earle from the *Narrative* (1839).

principal fort. It was something new to me to walk with Pistols &
Cutlass through the streets of a town. It has all ended in smoke. But
the consequence very disagreeable to us, since from the troubled state
of the country we cannot walk in the country.

Darwin's Diary recorded that, 'The revolutions in these countries are quite
laughable; some few years ago in Buenos Ayres, they had 14 revolutions in 12
months; things go as quietly as possible; both sides dislike the sight of blood;
& so that the one which appears the strongest gains the day.'

By the end of July and well into August Darwin did have the opportunity
to explore, and he records his experiences of hunting with John Clements
Wickham, Bartholomew James Sulivan, and a new midshipman by the
name of Robert Hamond. FitzRoy was delighted with his selection of
Darwin as his naturalist-companion. He wrote to Beaufort from
Montevideo on 15 August 1832:

All goes well – extremely well – on board. I can say, what seldom
may be said, with truth, that I do not wish to change a single Officer
or Man, and that I have not more sincere friends in the world than my
own Officers. From the Druid [HMS Druid] I have obtained an old
friend and shipmate named Hamond (a passed Midshipman) –
Captain Hamilton has lent him; and in some manner I must contrive
to keep him. Mr Darwin is a very superior young man, and the very
best (as far as I can judge) that could have been selected for the task.
He has a mixture of necessary qualities which makes him feel at
home, and happy, and makes every one his friend.

By 18 October the *Beagle* prepared to visit Tierra del Fuego to continue the
survey work and re-settle the Fuegians. Writing to Caroline while at sea,
Darwin relayed that,

...this second cruize will be a very long one; during it we settle the Fuegians & probably survey the Falkland Islands; After this is over (it is an aweful long time to talk about) we return to M Video, pick up our officers & then round the Horn & once more enter the glorious, delicious intertropical seas.

Darwin (from Montevideo on 24 November 1832) told Henslow of his expectations: 'I expect to find the wild mountainous country of Tierra del very interesting; & after the coast of Patagonia I shall thoroughly enjoy it'. The coast of Tierra del Fuego was reached on 16 December close to the southern part of Cape St Sebastian. The smoke from native fires was evidence of human habitation. On the 18th Darwin described in his Diary his first encounter with Fuegians in the Bay of Good Success:

When we landed, the party looked rather alarmed, but continued talking & making gestures with great rapidity. It was without exception the most curious & interesting spectacle I ever beheld. I would not have believed how entire the difference between savage & civilized man is. It is greater than between a wild & domesticated animal, in as much as in man there is greater power of improvement. The chief spokesman was old & appeared to be head of the family; the three others were young powerful men & about 6 feet high. From their dress &c &c they resembled the representations of Devils on the Stage, for instance in [Weber's opera] Der Freischutz.

Darwin continued:

The skin is a dirty copper colour. Reaching from ear to ear & including the upper lip, there was a broad red coloured band of paint; & parallel & above this, there was a white one; so that the eyebrows & eyelids were even thus coloured. The only garment was a large guanaco skin, with the hair outside. This was merely thrown over their shoulders, one arm & leg being bare; for any exercise they must be absolutely naked.

He thought, 'their very attitudes very abject, & the expression distrustful, surprised & startled'. Darwin recalled what Captain Cook had thought of their language:

Their language does not deserve to be called articulate: Capt Cook says it is like a man clearing his throat; to which may be added another very hoarse man trying to shout & a third encouraging a horse with that peculiar noise which is made in one side of the mouth.

FitzRoy's Fuegians were settled together at Woollya.

Every opportunity was made to undertake the extensive survey work in *Beagle* and the small boats. FitzRoy spent around two months engaged in this work in which Darwin assisted. On 26 February 1833 the *Beagle* sailed for the Falkland Islands, which were under possession of the British Crown, arriving on the morning of 1 March at Port Louis, the most eastern point of the islands.

'Button Island, near Woollya', engraving after Conrad Martens from the *Narrative* (1839). Button Island was named after Jemmy Button.

For Darwin the Falklands topography was

> *…remarkably easy to access to persons on foot; but half-concealed rivulets and numerous bogs, oblige a mounted traveller to be very cautious. There are no trees any where, but a small bush is plentiful in many vallies. Scarcely any views can be more dismal than that from the heights; moorland and black bog extend as far as eye can discern, intersected by innumerable streams, and pools of yellowish brown water.*

Darwin was beginning to ask some searching questions. He commented on the wild pigs and foxes of the Falklands. He was intrigued by the differences between the appearances of the foxes on the islands and those on the mainland of South America, especially in Patagonia. He believed there were radical differences. After visiting the islands he noted:

> *The only quadruped native to the island is a large wolf-like fox which is common to both East and West Falkland. I have no doubt it is a peculiar species, and confined to this archipelago; because many sealers, gauchos and Indians who have visited these islands, all maintain that no such animal is found in any part of South America…*

Early in March 1833 the sealing schooner *Unicorn* arrived in the Falkland Islands and FitzRoy capitalized on the owner's plight. The sealing season had been so unsuccessful that William Harris had run up huge debts. FitzRoy agreed to acquire the vessel to assist his survey work and thereby provide financial assistance to Harris. The vessel was renamed *Adventure* and sailed for the mainland of South America with Wickham in command on 4 April, followed two days later by the *Beagle*. They were making their way towards Maldonado in Uruguay where, according to FitzRoy, 'Mr Darwin lived on shore, sometimes at the village of Maldonado, sometimes making excursions into the country to a considerable distance'. FitzRoy occupied himself by working up his charts.

Writing to Catherine from Maldonado on 22 May 1833, Darwin made many requests for books, provisions and equipment in his postscript:

> *When you read this I am afraid you will think that I am like the Midshipman in Persuasion [Jane Austen], who never wrote home, excepting when he wanted to beg; it is chiefly for more books… Cary [the brothers John and William Cary were eminent scientific instrument makers with separate businesses in London] has 3s.6d tape measure of about 1½ feet. I have lost mine. I have at present a double convex lens [Darwin is referring to one of his microscopes], fitted to the object-glass, & about one inch in diameter; now I want one on a larger scale & with longer focal distance for illuminating opake [opaque] objects: it must be fixed on a stand & with plenty of motions…& lastly 4 pairs of very strong walking shoes from Howell.*

From Rio de la Plata on 18 July 1833 Darwin updated Henslow on his collection of specimens, and it was apparent he was looking forward to leaving the Atlantic coast of South America for the warmer waters of the Pacific:

> *After the Beagle returns…we take in 12 months provisions & in the beginning of October proceed to Tierra del F., then pass the Straights of Magellan & enter the glorious Pacific: The Beagle after proceeding to Conception or Valparaiso, will once more go Southward, (I however will not leave the warm weather) & upon her return we proceed up the coast, ultimately to cross the Pacific. I am in great doubt whether to remain at Valparaiso or Conception: at the latter beds of Coal & shells, but at the former I could cross & recross the grand chain of the Andes. I am ready to bound for joy at the thoughts of leaving this stupid, unpicturesque side of America.*

Darwin was also concerned about the fate of his specimens shipped home: 'I am anxious to know, what has become of a large collection (I fancy ill assorted) of Geological specimens made in former voyage from Tierra del Fuego'.

On 24 July the *Beagle* departed from the Rio de la Plata to complete her survey of the coast of Patagonia south of Bahia Blanca. Darwin landed on the north bank of the Rio Negro to explore, and on 11 August set off on horseback to visit the camp on the River Colorado of the Argentinian army who were fighting the Araucanian Indians, as well as other tribes and groups. He regaled Caroline with his expedition news claiming that he had 'become quite a Gaucho, drink my Mattee, & smoke my cigar, & then lie down & sleep as comfortably with the Heavens for a canopy as in a feather bed'.

Darwin was granted a meeting with General Juan Manuel de Rosas (1793–1877), the commander of the

'Patagonian', engraving after Phillip Parker King from the *Narrative* (1839). Darwin described the Patagonians as: 'Their height appears greater than it really is, from their large guanaco mantles, their long flowing hair, and general figure: on average their height is about six feet, with some men taller and only a few shorter; and the women are also tall; altogether they are certainly the tallest race which we anywhere saw'.

Argentinian forces who had accumulated vast wealth as a cattle farmer and beef exporter at the expense of the local people. Darwin was impressed by him but predicted the downfall of this South American dictator. In the early 1850s Rosas fled to Britain, where he spent the rest of his life in exile.

FitzRoy often wrote to Darwin if he was resident in one place for a significant period of time and his letter to Darwin of 4 October 1833 expressed concern at Darwin's escapades: 'How many times did you flee from the Indians? How many precipices did you fall over? How many bogs did you fall into? How often were you carried away by the floods?' In the same letter he informed Darwin of his appointment of Conrad Martens, his second shipboard artist.

Top: 'Santa Cruz River, and Distant View of the Andes', engraving after Conrad Martens from the *Narrative* (1839).

Bottom: Shooting guanacos on the banks of the Rio Santa Cruz, watercolour by Conrad Martens. This is a surprisingly rare portrayal of hunting by FitzRoy's second shipboard artist. Perhaps it was an activity that was regarded as commonplace and not worthy of recording. Martens himself was an excellent shot.

Darwin arrived in Montevideo in the first week of November 1833 to rejoin the *Beagle* but was surprised to find that the ship would not actually sail until the beginning of December. On 12 November Darwin was informing Henslow of yet more specimens to be shipped back. Among boxes of bones and stones, 'There are two boxes & a cask. One of the former is lined with a tin-plate & contains nearly 200 skins of birds & animals – amongst others a fine collection of the mice of S. America'.

Beagle finally left Montevideo on 6 December at 4 o'clock in the morning and after a protracted passage, mainly because the winds were 'light & foul', she arrived at Port Desire on the 23rd of that month. On Christmas Day FitzRoy organized sports and games and handed out prizes. Darwin participated in several land excursions. In the New Year, to speed up the official work, the *Beagle* and the *Adventure* separated. The former continued surveying in Tierra del Fuego. She entered the Strait of Magellan, and after a short stopover at Gregory Bay, arrived at Port Famine at the beginning of February 1834. The *Adventure* worked in the Falkland Islands.

At Woolaston Island Darwin went ashore and 'walked or rather crawled to the tops of some of the hills; the rock is not slate, & in consequence there are but few trees; the hills are very much broken & of fantastic shapes'. He was uncomplimentary about the appearance of six Fuegians: 'I never saw more miserable creatures; stunted in their growth, their hideous faces bedaubed with white paint & quite naked'. Darwin continued to ask searching questions, recording in his Diary on 24 February 1834: 'Whence have these people come? Have they remained in the same state since the creation of the world?' He continued: 'There can be no reason for supposing the race of Fuegians are decreasing, we may therefore be sure that he enjoys a sufficient share of happiness (whatever its kind may be) to render life worth having. Nature by making habit omnipotent, has fitted the Fuegian to the climate & production of his country'.

Before the *Beagle* sailed for warmer waters there was a second visit to the Falkland Islands. FitzRoy discovered that Argentinian prisoners had overpowered their guards and killed British residents. He was shocked to discover that Matthew Brisbane, who had been left in charge of the islands, had been brutally murdered. FitzRoy ensured that Brisbane was given a proper burial and succeeded in restoring order and British control.

Darwin was able to undertake some beneficial work: rocks were hammered, roots were pulled and he made extensive notes on the geological structure of the islands. He was of the opinion that the 'Zoology of the sea [was] generally the same here as in Tierra del Fuego'.

FitzRoy steered a course back to the Argentine coast and on 13 April *Beagle* dropped anchor at the mouth of the Rio Santa Cruz. FitzRoy had an ambitious plan that was unfulfilled from *Beagle*'s first expedition. He wanted to survey as far as possible up the Santa Cruz and he had small boats to assist him. The *Beagle* needed attention after she had hit a rock at Port Desire and FitzRoy thought the hull might be damaged. The beaching of *Beagle* to inspect her and make repairs resulted in an engraving after Conrad Martens' original drawing that featured in the *Narrative*. Martens also painted a wide-angle watercolour view showing three of the ship's boats on the Rio Santa Cruz that reveals the scale and challenge of the surveying enterprise. This picture was purchased from the artist later in the voyage by Darwin in Sydney. Darwin and FitzRoy both provided detailed

'Beagle laid ashore, River Santa Cruz', engraving after Conrad Martens from the *Narrative* (1839). FitzRoy wrote in the *Narrative*: 'On the 13th [April] 1834 we anchored in the Santa Cruz, and immediately prepared to lay our vessel ashore for a tide, to ascertain how much injury had been caused by the rock at Port Desire, and to examine the copper previous to her employment in the Pacific Ocean, where worms soon eat their way through unprotected planks'.

descriptions of the prevalent condors and guanacos of which many were shot. On 4 May the boats turned back from their arduous exploration of the river, which often involved the men actually hauling the boats by land lines. FitzRoy was concerned that supplies were running short. They had gone further than any other European but the men were frustrated at their progress.

Off the Chilean coast on Sunday, 20 July 1834, Darwin wrote to Catty:
When I wrote from the Falklands we were on the point of sailing for the S. Cruz on the coast of Patagonia. We there looked at Beagle's bottom; her false keel was found knocked off, but otherwise not damaged. When this was done, the Captain & 25 hands in three boats proceeded to follow up the course of the river S. Cruz. The expedition lasted three weeks; from want of provisions we failed reaching as far as was expected, but we were within 20 miles of [the] great snowy range of Cordilleras: a view which has never been seen by European eyes.

View of Mount Sarmiento, Tierra del Fuego, watercolour by Conrad Martens. Martens painted several views of this striking mountain. He borrowed a telescope from Darwin to help him create a picture for his sketchbook that he described as: 'Mount Sarmiento as seen from Port Famine by telescope distant 49 miles', dated 2 February 1834. It is now in the Cambridge University Library.

The river is a fine large body of water; it traverses wild desolate plains inhabited by scarcely anything but the Guanaco. We saw in one place smoke & tracks of the horses of a party of Indians: I am sorry we did not see them, they would have been out & out wild Gentlemen.

FitzRoy was vexed at the enforced sale of his auxiliary surveying vessel, *Adventure*, and realized that he could no longer make a complete survey of the Chilean and Peruvian coastlines. By the time Darwin had written his letter the *Beagle* had called in a Gregory Bay, spent a week at Port Famine and then sailed through and out of the Strait of Magellan, enjoying a spectacular view of 'the grand glacier' Mount Sarmiento en route. On 27 June 1834 she anchored at San Carlos at the northern end of the island of Chiloe. Darwin commented on the local people: 'They all appear to have a great mixture of Indian blood & widely differ from every other set of Spaniards in not being Gauchos'.

Beagle arrived at Valparaiso, Chile, on 23 July where she would be based for several months. Darwin made an extensive ride inland to Santiago. After Tierra del Fuego and Chiloe, Darwin found the climate much more to his liking: 'the sky so clear & blue, the air so dry & the sun so bright, that all nature seemed sparkling with life'. Darwin stayed in Santiago for a week where he was impressed by 'very pretty Signoritas' and observed that it was built 'on a plain, the basin of a former inland sea; the perfect levelness of this plain is contrasted in a strange & picturesque manner with great snow topped mountains which surround it'.

From Santiago Darwin travelled south 'about 40 leagues' to S. Fernando, from where he told Caroline: 'Every one in the city talked so much about the robbers & murderers'. He also told her that with the *Adventure* sold, 'we shall all be very badly off for room'.

When Darwin recovered from a protracted illness at Valparaiso he realized the extent of his captain's depression. Everyone now believed that the larger part of the expedition was over. FitzRoy was overworked and Darwin revealed in a letter that the 'difficulty of living on good terms with a Captain of a Man-of-War is much increased by its being almost mutinous to answer him as one would answer anyone else; and by the awe in which he is held – or was held in my time, by all on board'.

In an epistle to Catty written on 8 November 1834, Darwin conveyed his enthusiasm and envisioned the homeward leg:

SAN CARLOS DE CHILÓE.

SAN CARLOS DE CHILÓE.

Hurra Hurra it is fixed the Beagle shall not go one mile South of C. Tres Montes & from that point to Valparaiso will be finished in about five months. We shall examine the Chonos archipelago, entirely unknown & the curious inland sea behind Chiloe. For me it is glorious. C. T. Montes is the most southern point where there is much geological interest, as there the modern beds end...

The *Beagle* arrived in the harbour of San Carlos on Chiloe on 20 November where Darwin admired the volcanoes and observed the active motions and smoke of Osorno, one of the active volcanoes of the southern Chilean Andes, measuring 8,701 feet in height. In January 1835, from a distance, Darwin observed it erupting. Poor weather hindered *Beagle*'s survey work but she did establish that the island of Chiloe had been previously overestimated in terms of length by around 30 miles. Darwin collected frogs, flatworms and barnacles.

On 8 January 1835 the *Beagle* arrived at the predominantly Spanish-populated port and town of Valdivia. The ship was on her way north up the coast to Concepcion when the Great Earthquake occurred. The crew witnessed the tremendous swell of the sea passing along the western side of the Bay of Concepcion. Darwin was profoundly affected by the subsequent devastation and ruins that he saw, and Wickham sketched the remains of the cathedral. The *Beagle* survived but she lost all but one of her large anchors and in early March she reached to Valparaiso. Darwin embarked on an extensive overland expedition on horseback from Santiago to Mendoza. But he undertook his longest ride on 27 April 1835 through the Andes travelling more than 220 miles to Coquimbo (in search of new and remarkable natural history specimens and to observe the landscape). FitzRoy, travelling

'Remains of the Cathedral at Concepcion', engraving after John Clements Wickham from the *Narrative* (1839). On 4 March 1835 the *Beagle* entered the harbour of Concepcion and witnessed at first hand that there was barely a house left standing. Luckily nearly all the inhabitants escaped serious injury.

independently, met up with Darwin and painted a panoramic view of the town. Darwin summed up his frustrations and homesickness to Caroline: 'It is very hard & wearisome labour riding so much through such countries as Chili, & I was quite glad when my trip came to a close. Excluding the interest arising from Geology, such travelling would be down right Martyrdom'.

Darwin was more excited about his time in Peru. In mid-July the *Beagle* arrived at Iquique and later called into Callao, the port serving Lima. Syms Covington painted the striking ladies of Lima with their close-fitting outfits and covered faces that revealed only one large dark eye. The ladies and one of the fruits of the city made a lasting impression on Darwin:

> There are two things in Lima which all travellers have discussed; the ladies 'tapadas', or concealed in the saya y Manta, & a fruit called Chilimoya. To my mind the former is as beautiful as the latter is delicious. The close elastic gown fit's the figure closely & obliges the ladies to walk with small steps which they do very elegantly & display very white stockings & very pretty feet. They wear a black silk veil, which is fixed round the waist behind, is brought over the head, & held by the hands before the face, allowing one eye to remain uncovered. But then that one eye is so black & brilliant & has such powers of motion & expression that its effect is powerful.

Beagle's next port of call was the Galapagos Islands. She arrived there in mid-September 1835, and she was here for just five weeks. The ground-breaking discoveries stemming from this stopover would only emerge after the expedition was completed. Darwin had the opportunity to explore several areas of the island of Chatham, and also parts of the islands of Charles, Albemarle and James. 'Disgusting clumsy lizards' frequented the black lava rocks. Darwin established that they were vegetarian and could hold their breath under water for many minutes. He discovered new species of fish, observed and collected mocking birds and finches. The birds would be later identified and classified by the English ornithologist John Gould. The vital information he provided would lead Darwin to assert that the birds had evolved into different species on separate islands with individual shaped beaks for acquiring and processing food.

Land lizards and tortoises also caught Darwin's attention. Nicholas Lawson, an Englishman living on Charles Island and acting as governor, alleged that he could tell which island a tortoise came from by the shape of its shell; information that would later help Darwin construct his radical evolutionary theory relating to natural selection. The male tortoise was much larger than the female; 'some require 8 or 10 men to lift them'. They drank large quantities of water and wallowed in mud. Tortoises were killed for the water within them and for their meat. Many were taken onboard the *Beagle* for food to sustain the men on the rest of their journey home.

Beagle arrived at Tahiti in early November after a passage of more than 3,000 miles. Darwin did not much care for this '*fallen* paradise'. He lamented the nakedness of the populace and wished they would adopt a more European style of dress. Darwin was not only homesick, but by this stage of the expedition he had become noticeably jaded and his descriptions became less expansive and detailed. Even his chewing of the plant called *ava*, which had a powerful and hallucinogenic effect, failed to lift his spirits. The missionaries had forbidden the locals to use it and also encouraged temperance. Darwin visited Henry Nott, a veteran missionary of the island, who had recently completed a translation of the Bible into Tahitian.

Darwin hired a canoe to see something of the reef and admired the corals. It was his opinion that 'little is yet known, in spite of much that has been written, of the structure & origin of the Coral Islands & reefs'. He observed the diplomacy of the discussions between FitzRoy and Queen Pomare, when FitzRoy managed to extract an agreement of compensation of 36 tons of pearl oyster shells for the murder of British sailors that had occurred in her territorial waters before their arrival. Darwin described the Queen as 'an awkquard large woman, without any beauty, gracefulness or dignity of manners. She appears to have only one royal attribute, viz a perfect immovability of expression (& that generally rather a sulky one) under all circumstances'.

Beagle spent less than a month at Tahiti. Darwin did not think the island was of much interest. He would think even less of his time at the Bay of Islands in New Zealand, where the ship would stay for less than two weeks. However, he and FitzRoy were impressed with the missionaries there and satisfied themselves that the critical comments made about them by Augustus Earle were unfair. In Earle's book, *A Narrative of A Nine Month's Residence in New Zealand in 1827: Together With a Journal of a Residence in Tristan D'Acunha, an Island Situated Between South America and the Cape of Good Hope*, published in 1832, he claimed the missionaries were cold and inhospitable to him during his time there. However, after checking the facts, *Beagle*'s captain and naturalist, among many others, came to the conclusion that Earle was being economical with the truth.

Writing to Caroline from the Bay of Islands, on 27 December 1835, Darwin commented:

> *The Missionaries have done much in improving their [the Maoris'] moral character, & still more in teaching them arts of civilization…we are very indignant with Earle's book; besides extreme injustice it shows ingratitude. Those very Missionaries who are accused of coldness I know without doubt always treated him with far more civility than his open licentiousness could have given reason to expect…*

Darwin thought that the Maoris were superior in energy but that they were in every other way inferior to the Tahitians. In his opinion they were often unfriendly and warlike.

Sydney, Australia, was the last major stopover before the return to Britain. Darwin met up with his former shipmate, Conrad Martens, and he and FitzRoy purchased pictures from Martens that had been developed from his sketches produced during his time on the *Beagle* voyage. They also met Phillip Parker King, the commander of *Beagle*'s first expedition, who had

settled in Australia, and the ship's chronometers were verified at the Parramatta observatory.

Darwin explored the Blue Mountains on horseback, travelling to Bathurst. He rated the Aborigines only a few notches higher in terms of civilized beings to the Fuegians. He was impressed by Wentworth Falls, which was also painted by Martens. Darwin did not have much luck in observing kangaroos but he acquired and stuffed a duck-billed platypus. He pondered the bizarre creatures he had seen and recorded in his journal:

> *A little time before this, I had been lying on a sunny bank & was reflecting on the strange character of the Animals of this country as compared to the rest of the World. An unbeliever in everything beyond his own reason, might exclaim 'Surely two distinct Creators must have been at work; their object has however been the same & certainly the end in each case is complete'.*

Darwin was not sorry to leave Australia. He wrote in his Diary: 'you are a rising infant & doubtless some day will reign a great princess in the South; but you are too great & ambitious for affection, yet not great enough for respect; I leave your shores without sorrow or regret'.

To complete the chain of meridian observations required as part of the Admiralty Instructions, Captain Beaufort had given specific details of where the ship should make the necessary stopovers. Beagle proceeded to Hobart in Tasmania where she arrived on 5 February 1836. Darwin was able to examine

'View from the Summit of Mount York, New South Wales', watercolour by Augustus Earle, c. 1826–7. Earle noted that, 'the delinquents are employed in forming roads, by cutting through mountains, blasting rocks'. In the background of Earle's painting the Bathurst Plains are clearly visible. Darwin recorded that on 20 January 1836, 'In the afternoon we came in view of the downs of Bathurst. These undulating but nearly smooth plains are very remarkable in this country, from being absolutely destitute of trees. They support only a thin brown pasture'.

the geology of Tasmania and was given some specimens by Mr Frankland, the Surveyor General.

On 6 March the *Beagle* headed back to Australia, arriving at King George Sound close to the south-west corner of the mainland. They stayed for eight days and Darwin thought the scenery resembled that 'of the high sandstone platform of the Blue Mountains'. He was also very interested in the White Cockatoo Men, who 'were persuaded to hold a *corrobery*, or great dancing party'.

On 1 April *Beagle* arrived at the Cocos Islands in the middle of the Indian Ocean. These islands were dominated by coconuts, which was the main export product. Crabs had evolved to climb trees and open coconuts with their claws. Darwin was able to study the coral islands and later developed important ideas relating to their origin and structure. This resulted in his highly regarded publication *The Structure and Distribution of Coral Reefs* (1842).

Sir John Frederick William Herschel, 1st Baronet, albumen print, April 1867, by Julia Margaret Cameron. Sir John was the son of the astronomer Sir William Herschel, a science in which he too would excel. He was also a mathematician, chemist and experimental photographer who encouraged Cameron in her work.

The last ports of call according to Admiralty Instructions were at Mauritius, the Cape of Good Hope and Ascension Island. Darwin's homesickness was evident in his letter to Caroline on 29 April 1836:

Now one glimpse of my dear home would be better than the united kingdoms of all the glorious Tropics. Whilst we are at sea, & the weather is fine, my time passes smoothly, because I am very busy. My occupation consists in rearranging old geological notes: the rearranging generally consists in totally rewriting them. I am just now beginning to discover the difficulty of expressing one's ideas on paper. As long as it consists solely of description it is pretty easy; but where reasoning comes into play, to make a proper connection, a clearness & a moderate fluency, is to me, as I have said, a difficulty of which I had no idea.

On the first day of May Darwin climbed the La Pouce mountain of 2,600 feet on the south side of Port Louis, Mauritius. He and John Lort Stokes visited Captain Lloyd, the surveyor general, and stayed at his delightful country residence enjoying a ride on his elephant. Describing the ride, Darwin wrote: 'The circumstance which surprised me most was its quite noiseless step'. By the end of the month *Beagle* arrived in Simon's Bay at Cape Town, where Darwin was assisted by Dr Andrew Smith, a military surgeon and leading authority on the zoology of South Africa. He also met the English mathematician, astronomer, chemist, artist and experimental photographer, Sir John Herschel (1792–1871) who was making a systematic study of the southern skies through his 20-foot telescope.

Herschel was a brilliant but shy man, and Darwin was motivated and indebted to this remarkable polymath. In his *Autobiography* Darwin wrote that in addition to Humboldt's *Personal Narrative*, 'Sir J. Herschel's *Introduction to the Study of Natural Philosophy* stirred up in me a burning zeal to add even the most humble contribution to the noble structure of Nature Science'.

Both Herschel and Darwin were sensitively photographed by Julia Margaret Cameron (1815–1879), who excelled at capturing celebrities, including actors, artists, musicians, poets and scientists. Darwin declared her pensive profiles of him to be the finest he had ever seen and delighted in her company when he and his family visited her in 1868 at Freshwater on the Isle of Wight during a holiday. In the opening lines of Darwin's *On the Origin of Species* he was referring to Herschel when he wrote that his intent is 'to throw some light on the origin of species – that mystery of mysteries, as it has been called by one of our greatest philosophers'.

By 8 July the *Beagle* had arrived at St Helena where Napoleon's tomb and dilapidated house were visited. The house now sold wine and there was a billiard table to amuse the visitors. Darwin's assistant made a drawing of what everyone regarded as a very simple memorial for such a remarkable politician and military leader. Darwin was delighted to discover a species of dung-eating beetle.

After five days of sailing *Beagle* reached Ascension Island. There was little that could be described beautiful about these roughly triangular-shaped volcanic rocks. The islands were sparsely populated with British marines and 'Negroes liberated from slave ships'. Darwin devoted himself to writing up his notes and was counting the days till the ship would set sail for home. But before the *Beagle* arrived at Falmouth FitzRoy wanted to double-check the meridian distance that he had measured at Bahia earlier in the expedition and the ship crossed the Atlantic to the Brazilian coast once more. Darwin had lost none of his passion for the tropical scenery:

> *In the last walk I took, I stopped again and again to gaze on such beauties, & tried to fix for ever in my mind, an impression which at the time I knew must sooner or later fade away. The forms of the Orange tree, the Cocoa nut, the Palms, the Mango, the Banana, will remain clear & separate, but the thousand beauties which unite them all into one perfect scene, must perish; yet they will leave, like a tale heard in childhood, a picture full of indistinct, but beautiful figures.*

The *Beagle* started for home on 6 August and this time there was no crossing-the-line ceremony. A brief stopover at Port Praya in the Cape Verde Islands failed to live up to Darwin's first impressions on the outward voyage. The last stopover was made at the Azores where Darwin noted with extreme brevity, 'we stayed for six days'. Finally, on Sunday, 2 October 1836 *Beagle* dropped anchor at Falmouth, Cornwall, in England.

Darwin made a hasty exit from the ship and travelled by coach as fast as he possibly could to his family home in Shrewsbury. The *Beagle* would continue on to Greenwich for a final verification of her shipboard instruments before being paid off at Woolwich – the place of her birth.

Chapter 5
The Art of Surveying the Sea

It is my duty to inform you of the excessive exertions made by Mr John Lort Stokes – Mate, and Assistant surveyor; and to request that you will represent his conduct to the Lords Commissioners of the Admiralty, in the manner which you think proper. The greater part of the accompanying charts, and other documents, are his work, and I regret to say that his incessant occupation has brought upon a serious pectoral affection. – Captain Robert FitzRoy, writing to Captain Francis Beaufort, from Monte Video, 30 November 1833

'Officers Surveying', watercolour by Captain Owen Stanley. Stanley portrays officers engaged in surveying from the land with various instruments. In the centre a man looks through a theodolite, and to the left a seaman crouches as he makes observations through a sextant on a fixed stand. The officer beside the tent is studying a dip circle, used for measuring magnetism.

Captain Robert FitzRoy had a dual role onboard HMS *Beagle*. In addition to commanding the ship he was also the senior surveyor. John Lort Stokes worked alongside him as his assistant and ship's mate.

FitzRoy had received specialist training to create and supervise the production of the most accurate coastal profiles and views, maps and plans, for the compilation of printed Admiralty sea charts. Sea charts could be described as specific, and usually detailed, maps of the sea. The preparatory work for the charts took place on the large drafting table that

dominated Darwin's cabin. By his own admission Darwin had no drawing skills but he would have observed the work being undertaken. Unfortunately he did not write about the men's drafting duties.

By the early 1800s large areas of South American waters remained uncharted. Substantial diplomatic benefits and trade opportunities could be gained by those countries who created, or acquired, the charts. Old Spanish charts existed for the purpose of supplying and maintaining her colonies, but few were freely available to other Europeans and the majority were inaccurate.

From the end of the Napoleonic Wars in 1815 Britain started a concerted campaign to commission ships to investigate coastlines, inshore waters and rivers across the globe. *Beagle* was part of the strategy to establish an influential British presence in South America. Data was collected relating to the depth and nature of the sea or riverbeds. Sandbanks, shoals and reefs were identified and records made of currents, tides and winds. To collate this information soundings were taken, and observations were made with scientific instruments to obtain measurements afloat and from the shore.

'Sounding Winch', graphite drawing by Captain Owen Stanley. This winch was vital for establishing the depths of coastlines. This drawing, and the 'Officers Surveying', form part of an album of sketches compiled by Stanley during his command of the surveying ship, HMS *Rattlesnake*, 1846–9. Stanley was tasked to survey the inner route along the Great Barrier Reef and chart the southern coast of New Guinea.

Surveying parties often made coastal landings to take the various measurements and angles required for the survey. Captain Owen Stanley of the *Rattlesnake* created watercolours and drawings showing the survey parties at work using a variety of instruments. He also drew a picture of a type of newly invented deep-sea sounding winch that *Beagle* had onboard. This was used to take soundings to measure the depth of the ocean floor: from the small boats in shallower water the men would have used the traditional sounding or lead line. FitzRoy was laudatory of the benefits of the winch writing in the *Narrative*:

> *In again trying for soundings with three hundred fathoms of line, near the Island of St. Jago, we became fully convinced of the utility of a reel, which Captain Beaufort had advised me to procure. Two men were able to take in the deep sea line, by this machine, without inter-*

fering with any part of the deck, except the place near the stern, where the reel was firmly secured. Throughout our voyage this simple contrivance answered its object extremely well, and saved the crew a great deal of harassing work.

The surveyors would have been equipped with sighting telescopes and theodolites. Sextants were used to establish latitude, a ship's distance north or south of the equator (0°), by astronomical observations. Longitude, a ship's distance east or west on the Earth's surface from a Prime Meridian (0°), was determined by chronometers. Thanks to FitzRoy, *Beagle* had an ample supply of these onboard, which enabled him to establish longitude by comparing the local time to the time at the Greenwich Meridian, used by British shipping as 0° Longitude for navigation and surveying. (Since the Washington Conference of 1884, Greenwich has been used internationally as the world's Prime Meridian.)

To create accurate measurements the men used the established system of triangulation. The triangulation process required accurate knowledge of the distance between two of three points. Taking angular measurements from these enabled the third point to be precisely determined using theodolites or sextants. All of this data was vital to create an accurate chart.

FitzRoy noted the challenges of the survey work in his Appendix (No. 39, Notes on Surveying a Wild Coast) to the *Narrative*:

Our first object was to find a safe harbour in which to secure the ship. There we made observations of latitude, time and true bearing; on the tides and magnetism. We also made a plan of the harbour and its environs; and triangulations, including all the visible heights, and more remarkable features of the coast, so far as it could be clearly distinguished from the summits of the highest hills near the harbour. Upon these summits a good theodolite was used...

It was also essential for the seafarer to be able to recognize geographical locations. So in addition to the scientific data, the surveyors and other officers with artistic ability would create sketches, drawings and watercolours of ports, harbours and landfalls, headlands, hills and mountains, and any other readily identifiable topographical points of interests. Information relating to safe anchorage and the availability of water and wood was recorded. Collectively this information would feature on the charts to aid navigation.

In 1795 the British Admiralty established a specialist department, the Hydrographic Office, for the study of the surface waters of the earth and arrange the printing of the charts. Initially based in Whitehall, London, the Office is now in Taunton, Somerset, and is the world's largest publisher of nautical charts; proudly boasting that it has been 'charting the world for over 200 years'.

The first Admiralty Hydrographer was Alexander Dalrymple (1737–1808). He was appointed by order of George III in 1795. He set about reviewing the 'difficulties and dangers to his Majesty's fleet in the navigation of ships'.

The work rate of producing charts was initially very slow and the first Admiralty chart of Quiberon Bay in Brittany was not published until 1800.

Originally from Edinburgh, Dalrymple was an overly ambitious man with an argumentative temperament. He had gained seafaring and administrative experience in the East India Company, with whom he argued and fell out. However, some of his achievements were recognized as significant, and he was elected a Fellow of the Royal Society in 1765.

Dalrymple aggressively canvassed the Admiralty to be given charge of what in fact became Lieutenant James Cook's first voyage to the Pacific in HM Bark *Endeavour*. Dalrymple was bitterly disappointed: the Admiralty were insistent that only a naval officer could take command of one of their ships. Dalrymple was adamant that the so-called Southern Continent existed and could be exploited for King and Country. His ardent beliefs featured in his *An Account of the Discoveries Made in The South Pacifick Ocean, Previous to 1764*, published in 1767. Since ancient times it was a widely held belief that there must be a large temperate land mass in the far south to balance the land masses in the north. Cook's first and second voyages of discovery, ranging over the years 1768 to 1775, proved that no such land mass existed.

Cook took a copy of Dalrymple's publication with him on his first voyage and found parts of it of great use, notably the inclusion of details of a sea-route between North Australia and New Guinea, a route discovered by the Spanish seaman Luiz Vaez de Torres in the early 1600s. Dalrymple came across

'The Arctic Council', oil painting by Stephen Pearce (1819–1904). The Arctic Council was not a formal organization, but the people were real. Seated in the centre is Captain Francis Beaufort. The 'Council' provided advice during the search for Sir John Franklin who disappeared searching for the North-West Passage. Some of the 'Members', including Frederick William Beechey, Sir James Clark Ross and Sir Edward Sabine, also had experience of sailing in southern waters.

Torres' report of this voyage in Manila in 1769 and he was the first to name it the Torres Strait.

Dalrymple wrote another publication, *An Historical Collection of the Several Voyages and Discoveries in the South Pacific Ocean* (London, 1770), in which he continued to argue the case for the existence of a Southern Continent. Dalrymple's cantankerous temperament was his undoing. On 28 May 1808 he was dismissed from the office of Hydrographer and in the following month he died, according to medical attendants, of 'vexation'.

Captain Thomas Hannaford Hurd was appointed as the second Hydrographer from 1808 until his death in 1823. Born in 1747 in Plymouth he joined the navy on 1 September 1768 as an able seaman. He was of a more even temperament than his predecessor and a naval officer who had served with distinction under Admiral Lord Howe (1726–1799). He was selected for his past surveying experience (in 1789 the Admiralty had sent Hurd to make a detailed survey of Bermuda).

Hurd was given permission to sell charts to the mercantile marine and to the general public. In 1809 he was also given responsibility for the supply of timekeepers, and oversaw the production of volumes of sailing directions and the first chart catalogue.

Hurd was followed by Sir William Edward Parry (1790–1855), who was renowned as a polar explorer, although he was only in the position for six years from 1823 to 1829, covering *Beagle*'s first survey expedition.

Captain Francis Beaufort (1774–1857) was the fourth Hydrographer, and the one most closely associated with FitzRoy and the *Beagle*. He held the position for more than twenty-five years from 1829 to 1855, and was one of the most hardworking, sympathetic and supportive naval officers and administrators of his era. Originally from Navan in County Meath, Ireland, Beaufort's father was the rector of Navan and a topographer and architect of note.

Beaufort became a gifted inventor: among his many accomplishments is the Beaufort Wind Scale, developed in 1805 to gauge wind strength. FitzRoy was tasked to test it out on the second voyage. He also introduced official Tide Tables in 1833 and in the following year Notices to Mariners, which were published updates of what the Hydrographic Office now calls 'safety-critical navigational information', that were used in conjunction with older charts until they were updated. By 1855, the year of Beaufort's retirement, the Hydrographic Office's chart catalogue listed 1,981 charts, with 64,000 copies issued to the Fleet.

Thanks to the hard work and vision of Beaufort the Admiralty became the foremost authority in the world in the techniques of hydrography and cartography. He had personal first-hand experience of undertaking survey work in South American waters. He supervised surveys of the Mediterranean, the African coast, the Pacific Ocean, South America, Australia, New Zealand

and China, and various polar voyages. Among the long list of eminent naval men trained and guided by Beaufort are the names of Frederick William Beechey, Edward Belcher, Henry Foster, Phillip Parker King, Henry Charles Otter, William George Skyring, William Henry Smyth, Owen Stanley, Alexander Vidal, and Robert FitzRoy. Beaufort became an admiral and was knighted for his naval and hydrographic services.

FitzRoy and Beaufort maintained a fairly regular and revealing official and private correspondence until the latter's death in 1857. In fact FitzRoy's Admiralty Instructions categorically stated he 'should keep up a detailed correspondence by every opportunity with the Hydrographer'.

In part, Beaufort was also responsible for the inclusion of someone who later became the most celebrated member of *Beagle*'s crew. It was through his contacts that Darwin was introduced to FitzRoy. Jordan Goodman, author of *The Rattlesnake – a Voyage of Discovery to the Coral Sea*, one of the expeditions commanded by Owen Stanley, aptly summarizes Beaufort's objectives and the importance of carrying naturalists aboard survey ships:

> *He thought of surveying in a broad sense of scientific investigation and knowledge. To him, hydrography, was not just a matter of knowing coastlines…and so forth…but also included natural history, both of land and sea, geography and ethnography. To this end, he wanted survey ships to carry naturalists to make the necessary observations and collections to further that branch of knowledge and to fill the metropolitan, provincial and private museums with the exotic specimens of flora and fauna they encountered. He wanted discovery and explorations to be at the centre of marine surveys.*

Beaufort and FitzRoy shared many common interests and personal battles with the Admiralty. The Hydrographer was a founder Member of the Royal Geographical Society in 1830. Seven years later, as a result of his successful survey work on the *Beagle*, FitzRoy was awarded the Society's Gold Medal. Whenever possible Beaufort tried his best to help his colleague, although he was not always able to persuade the Admiralty in his favour.

On 26 March 1833 FitzRoy purchased the schooner *Unicorn* in the Falkland Islands, because he was certain that he could speed up the survey work if he acquired an additional vessel. She had been built as a private yacht and later armed and used by Lord Thomas Cochrane, the inspiration for several fictional heroes including Patrick O'Brian's Captain Jack Aubrey. It was love at first sight, Fitztroy wrote: 'A fitter vessel I could hardly have met with, one hundred and seventy tons burthen, oak built and copper fastened throughout, very roomy, a good sailer, extremely handy, and a first-rate sea-boat'.

FitzRoy used $6,000 (nearly £1,300) of his own money, as he did not have Admiralty approval to hire or purchase additional vessels, and paid considerably more to fit her out. She was re-named *Adventure*, the name of the main escort ship on *Beagle*'s first voyage, and according to Darwin, in deference to the ship that sailed with Captain Cook's *Resolution* on his second voyage of discovery. FitzRoy wrote to Beaufort in a confident, optimistic tone: 'I believe that their Lordships will approve of what I have done…but if I am wrong, no inconvenience will result to the public service, since I alone am responsible for the agreement…and am able and willing to pay the stipulated sum'. FitzRoy had also written directly to the Admiralty. The secretary at the Admiralty had underlined the words 'am able and willing' and wanted Beaufort's opinion on the matter. Beaufort placed on record:

> There is no expression in the Sailing Orders, or surveying instructions, given to Commander FitzRoy which convey to him any authority for hiring and employing any vessels whatever. On the other hand, there can be no doubt that by the aid of small craft he will be sooner and better able to accomplish the great length of coast which he has to examine – and which seems to contain so many unknown and valuable harbours; – especially if he finds it necessary to trace the course of a great river, which had been reported to him as being navigable almost to the other side of America. It may also be stated to their Lordships that the Beagle is the only surveying ship to which a smaller vessel or Tender has not been attached.

Despite Beaufort supporting his case, the Admiralty's response was negative. FitzRoy argued his case in further correspondence, but he was ordered to sell the *Adventure* on 22 July 1834: the Admiralty was adamant that he should focus his command solely on the *Beagle*. The sale was, in his own words, 'very ill-managed, partly owing to my being dispirited and careless'. He sold the vessel for just £1,400: the extra costs of fitting her out meant that he made a significant loss in the enterprise. Early in the voyage, in October 1833, FitzRoy had hired two considerably smaller vessels, the *Paz* (15 tons) and the *Liebre* (9 tons) to assist the survey work. They were relinquished in August 1834. FitzRoy personally paid the hire cost of £1,680. Stokes recalled that, 'These craft did one or other of the officers survey the coast from the Rio Plata to the Straights of Magellan over a period of nearly twelve months, whilst the Beagle was engaged further south'.

FitzRoy had transformed the small boats:

> No person who had only seen the Paz and Liebre in their former wretched condition, would easily have recognised them after being refitted, and having indeed almost new equipment. Spars altered, and improved rigging, well-cut sails, fresh paint, and thorough cleanliness, had transformed the dirty sealing craft into smart little cock-boats: and as they sailed out of Port Belgrano with the Beagle, their appearance and behaviour were by no means discouraging.

Perhaps because the survey work was routine rather than something extraordinary there are remarkably few expansive and detailed descriptions. FitzRoy wrote one of the most in-depth to Beaufort from Maldonado, 7 June 1833: 'I cannot omit an opportunity of telling you how we are going on, though I have no "Documents" ready for you yet'. He continued:

> In those cockleshells [the Paz and Liebre, called 'cockleshells' because they were small vessels] – the Coast between Port Desire and

Bahia Blanca has been explored satisfactorily, and when you see the New Charts, you will say 'I had no Idea there remained so much to be done on that Coast.' Their work has been confined to the immediate vicinity of the land; there is still work for the Beagle near tide races and outlying shoals. They have found a new River called by the Indians 'Chubat' in Lat. 43 degrees 20' S. Long. 65 degrees 15' W., which though not large nor deep, is rapid & flows through a most fertile country with so winding a course that a succession of Islands, or 'water meadows', might be formed by canals, thus which would be at the same time fortified against the Indians who never cross water in their attacks upon civilized man. Lt. Wickham says the river is about 100 yards wide, and will admit vessels of thirty tons burthen.

FitzRoy reaffirmed the use to Beaufort of the *Paz* and *Liebre*:

The Labyrinth between San Blas & Bahia Blanca is partly finished. Mr Stokes is now working at that part. Off the 'Islas de la Confusion', the aforesaid Labyrinth, there are shoals, out of sight of land, & extremely dangerous. In the Beagle, although so small, we could not have overhauled these places well, nor half so quickly, as the cockleshells. They are but decked boats...but their work will speak for them.

In *Life and Letters* Bartholomew James Sulivan has left behind some of the most insightful observations of the survey work in the *Paz* and *Liebre*, including descriptions of their interiors, and the activities of ship's boats:

The cabin in Stokes' craft [Liebre] is seven feet long, seven wide, and thirty inches high. In this three of them stow their hammocks, which in the daytime form seats and serve for a table. In a little space forward, not so large, are stowed five men. The larger boat carries the instruments. Her cabin is the same size, but is four feet high, and has a table and seats.

In December 1834 Sulivan participated in a separate survey in one of *Beagle*'s boats of the east side of the island of Chiloe and islets in the Gulf of Ancud, a large body of water separating Chiloe Island from the mainland of Chile. He was assisted by Darwin, three officers and ten men. He revealed

'Bivouac at the head of Port Desire inlet', after Conrad Martens, engraved as an illustration for the *Narrative* (1839). This plate also includes three other views of Port Desire. Richard Darwin Keynes describes his great-grandfather's explorations there at Christmas-time 1833. He 'took a long walk to the north, where he found a veritable desert composed of gravel, with rather little vegetation and not a drop of water. The dryness of the country was nevertheless redeemed by the fact that it supported many guanacos'. The uppermost illustration reveals that there were plenty of rabbits.

some of the lighter moments amidst the essential but mundane task of surveying by day and night: 'We are all then crowded into one tent and went on singing till twelve, and I have never under any circumstances saw a more merry party. All the comic songs that any one knew were mixed up yarns of English, Scotch and Welsh'.

FitzRoy thought very highly of Sulivan. He was successful in recommending him for a later survey voyage. Writing after the end of his *Beagle* command from his land-based address at 31 Chester Street, FitzRoy gives a carefully balanced account of the strengths and weaknesses of his former officer but concludes that he 'is *up to the business completely*. He is as thorough a seaman, for his age, as I know…. He is an excellent observer, calculator, and surveyor'. However, FitzRoy did note that, 'He is not a *neat* draftsman, though his chart work is *extremely* correct. (His hand is not quick enough for his mind, or his mind is too quick for his hand)'.

The official Admiralty Instructions, compiled largely by Beaufort, offered guidance on how *Beagle*'s surveyors should approach the task of compiling coastal profiles. Officers were not to waste time creating artworks. There was an acknowledgment of the significant workload of each officer:

In such multiplied employments as must fall to the share of each officer, there will be no time to waste on elaborate drawings. Plain, distinct roughs, every where accompanied by explanatory notes and on a sufficiently large scale to show the minutiae of whatever knowledge has been required, will be documents of far greater value in this office, to be reduced or referred to, than highly finished plans, where accuracy is often sacrificed to beauty.

The Admiralty was clearly of the opinion that too much time was wasted on drawing hills:

…which in general cost so much labour, and are so often put in from fancy or from memory after the lapse of months, if not years, instead of being projected while fresh in the mind, or while any inconsistencies or errors may be rectified on the spot. A few strokes of the pen will denote the extent and direction of the several slopes much more than the brush, and if not worked up to make a picture, will really cost as little or less time. The in-shore sides of hills, which cannot be seen from any of the stations, must always be mere guess-work, and should not be shown at all.

South West Opening of Cockburn Channel (Mount Skyring), engraving, after W. W. Wilson from the *Narrative* (1839). Although only the name Wilson is inscribed below this plate (he served on HMS *Adventure* during *Beagle*'s first expedition) other officers and artists, including Martens, almost certainly contributed to this multi-view sheet.

Beagle's surveyors had to provide the perpendicular height of all remarkable hills and headlands and,

All charts and plans should be accompanied by views of the land; those which are to be attached to the former should be taken at such a distance as will enable a stranger to recognize the land, or to steer for a certain point; and those best suited for the plan of a port should show the marks for avoiding dangers, for taking a leading course, or choosing an advantageous berth. In all cases the angular distances and the angular altitudes of the principal objects should be inserted in all degrees and minutes on each of the views, by which they can be projected by scale so as to correct any want of precision in the eye of the draftsman. Such views cannot be too numerous; they cost but a few moments, and are extremely satisfactory to all navigators.

'Southern Portion of South America', map from the *Narrative* (1839) engraved by J. Dower, Pentonville, London. Dower was a London-based draughtsman, engraver and publisher.

The Admiralty was not paying FitzRoy's team to paint pretty pictures. They wanted the official surveyors to produce images to improve navigation. To this end the creation of coastal profiles and views was a vital part of the production process to create accurate Admiralty charts.

The importance of creating views of the coastlines from the sea had been highlighted since the Admiralty Instructions of 1759. They contained a long list of things that were considered important to observe and record. These observations were to include: sands, shoals, sea marks, soundings, bays and harbours, times of high water and setting of tides, and, in particular, information such as the best anchoring and watering places, and descriptions of the best methods of obtaining water, fuel, refreshment and provisions. Fortifications were also to be described and their form, strength and position noted. The Instructions specifically indicated that if a ship carried any capable artists they should provide illustrations, with references and explanations attached, of this information. The material should then be submitted to the Admiralty in London.

Naming newly encountered places was a challenge for all explorers and surveyors. *Beagle*'s Admiralty Instructions (drawn up by Beaufort) offered advice and guidance on how to proceed:

> *Trifling as it may appear, the love of giving a multiplicity of new and unmeaning names tends to confuse our geographical knowledge. The name stamped upon a place by the first discoverer should be held sacred by the common consent of all nations; and in new discoveries it would be far more beneficial to make the name convey some idea of the nature of the place; or if it be inhabited, to adopt the native appellation, than to exhaust the catalogue of public characters or private friends at home. The officers, and crews, indeed, have some claim on such distinction.*

Indeed, the Instructions went so far as to encourage the use of names of officers and crew, 'which slight as it is, helps to excite an interest in the voyage'.

Completed charts, plans, views, drawings and watercolours, as well as letters and Darwin's specimens, were sent back to England at every available opportunity that *Beagle* came into contact with a British naval or merchant ship on route home. Writing to Beaufort in a private letter from Valparaiso on 14 August 1834, FitzRoy confides:

> *A merchantman going to Liverpool tempts me to send you a short letter…as there is much material I doubt it will be ready for the Samarang [HMS Samarang]. I shall try hard. The Falkland survey is better than I had reason to expect – all the exterior is well laid down. There is much to do about the eastern entrance of the Straits of Magellan – extensive banks and very difficult tides. I do not think the present copper plates of Magellan's Strait will allow of the new work being combined with the old – a new edition will perhaps be required. The title of the present charts stands in the way of an extensive and well sheltered port on the east coast of Tierra del Fuego.*

"No 17" (22)

His Majesty's Surveying Sloop "Beagle," Valparaiso 2d Nov 1834.

Sir,

Herewith I have the honor of forwarding to you the Charts – Views – and Directions – mentioned in the enclosed list: (see Note)

I beg to call your attention to the Chart of the Falkland Islands, the work of Lieutenant J. C. Wickham and Mr. A. B. Usborne, Master's Assistant, and to solicit you to notice the merit and inferior station of the latter; a statement of whose services I enclose.

I likewise send a statement of the services of Mr. John Lort Stokes – Mate and Assistant Surveyor – whose

Note. The List is inside the tin case which contains the Charts. R.F.

Captain Beaufort R.N.

whose exertions have been unceasing – as the charts sent home will shew.

I have the honor of being
Sir
Your very obedient
humble Servant
Robt FitzRoy Commander

Robert FitzRoy's letter to Captain Francis Beaufort from Valparaiso, 2 November 1834, informing the Hydrographer that the list of charts, views and directions is inside the tin case that contains the charts.

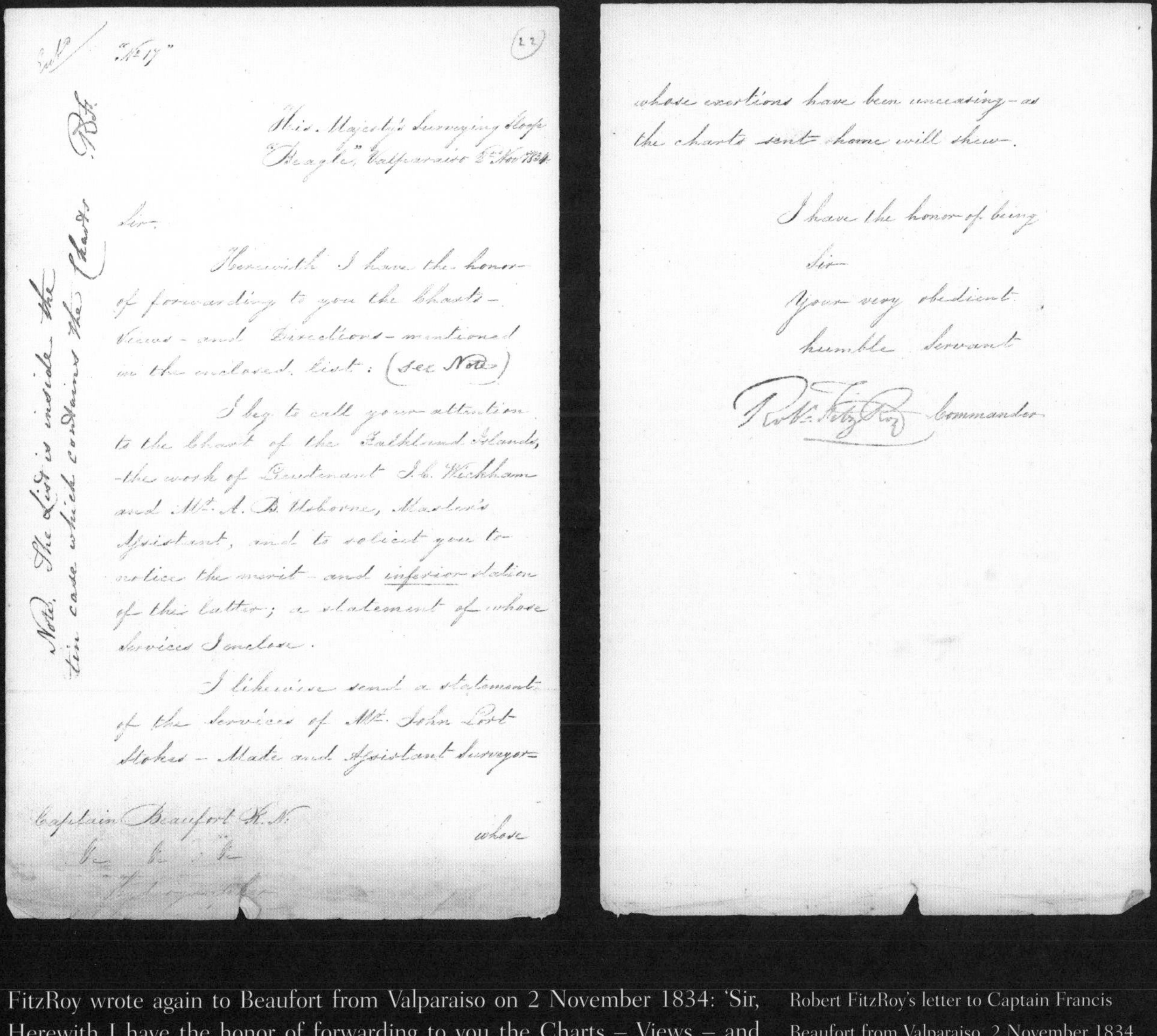

FitzRoy wrote again to Beaufort from Valparaiso on 2 November 1834: 'Sir, Herewith I have the honor of forwarding to you the Charts – Views – and Directions – mentioned in the enclosed list'. And, 'Note. The list is inside the tin case which contains the charts'. Included in this consignment were twenty-three charts including the east and west coast of the Falklands, Strait of Magellan and Beagle Channel, and ninety-five views including the interior of Patagonia and Tierra del Fuego. Also ten plans including Middle Cove, part of the north side of Wollaston Island and part of Hardy Peninsular, Port William and Port Desire.

During the voyage FitzRoy was also receiving consignments of reference material from Beaufort. Two years previously on 16 August 1832 FitzRoy wrote to him requesting he be,

> *...furnished with the following Charts and Papers, as they will expedite the work very materially.*
>
> *Charts*
> *River Plate, two sheets, two Copies*
> *Captain King's Charts of these Coasts (when published), six Copies*
> *South Sea, in three sheets, two Copies*

Tracing Paper…two squires
The two copies of the River Plate, two copies of Captain King's Charts,
and two of the South Sea, are for working upon, for laying down every
addition which is in our power to make, and sending them to England
at different times without depriving ourselves of the only copy on board.
The tracing paper is extremely useful. Having a large glass tracing
frame, no opportunity of tracing Charts or Plans, printed or manu-
script, which I can borrow or buy, and which promise to be of any use,
is omitted; and I did not bring enough from England.
The other four Copies of Captain King's Charts are requested by me
for the sole purpose of giving them away from time to time to the
Local Authorities who may render us assistance in our progress, and
who did materially assist the Adventure and the Beagle in their former
voyage. May I beg also that the Charts and Paper may be packed in tin
and rolled instead of doubled.

If FitzRoy believed that he might have a discrepancy in his chart work measurements or observations, he would retrace his steps until he ascertained the true results. He proved that the French admiral and explorer Baron Albin Rene Roussin (1781–1854) had made errors in his charts created during his 1819 exploration of the coasts of Brazil and the Abrolhos Islands. He wrote to inform Roussin of his findings but received no reply.

An open folio containing coastal profiles and views can be seen in an oil painting created by Augustus Earle, FitzRoy's first artist. Earle's picture

'Life on the ocean…', oil painting by Augustus Earle. The deliberate close-cropping of Earle's compositions and visually striking juxtaposition of objects with people allies the artist to the genre of caricature. Note the curious view of legs on the steps (upper left) and the marine's head and upper shoulders emerging from the hatchway. But Earle also succeeds in capturing the camaraderie and cooperative activities of life afloat.

portrays the interior of a midshipmen's berth at sea and the folio is clearly visible on the deck. Some naval instructors are teaching various disciplines: as part of their on-going naval training midshipmen (trainee officers) were taught by their superiors to draw and use watercolour. The painting also portrays navigational instruments including sextants and a telescope.

In Earle's quirky composition he plays around with conventional scale and perspective. He exaggerates the height between the decks. It is likely that this was the painting exhibited at the Royal Academy of Arts in 1837 under the title 'Life on the ocean, representing the usual occupation in the steerage of the young officers of a British frigate at sea'.

Earle's painting was produced in the penultimate year of his life with a companion picture entitled 'Divine Service as it is Usually Performed on Board a British Frigate at Sea' (also exhibited at the Royal Academy in 1837). It has been suggested by some writers (including Dr Janet Browne) that this picture portrays Darwin, FitzRoy (as the captain reading the Bible) and Fuegia Basket. This is wishful thinking. The pair of pictures certainly derive from Earle's first-hand seafaring experiences, and more specifically relate to his time aboard HMS *Hyperion*. Earle's oil paintings are now at the National Maritime Museum, Greenwich, and the related preparatory watercolours are part of the Rex Nan Kivell Collection, National Library of Australia in Canberra. The watercolours contain considerably less detail, although the height of the ship's decks is more authentic than the vast spaces portrayed in

'Divine Service…', oil painting by Augustus Earle, almost certainly the companion picture of the painting opposite, exhibited at the Royal Academy of Arts in 1837. If this does portray Admiral Searle, his appearance certainly differs from his description in the *Oxford Dictionary of National Biography*, which describes him as being a 'handsome, strongly built man of middle height, with black hair and dark complexion'.

the final oil paintings. One of the watercolours has a title (originating from Earle) that indicates the date '1820' and the name of the ship *Hyperion* inscribed on the mount: '130 Divine Service on board a British Frigate, H.M.S. Hyperion 1820'. Assuming that Earle's pictures feature accurate portraits rather than tableaux, the captain of this ship at that time would have been Admiral Thomas Searle (1777–1849). Earle sailed on this ship shortly after the start of the blockade by Lord Thomas Cochrane, of Callao, the largest and most prominent port in Peru, in November 1820, during the second phase of Chile's War of Independence 1818–26, during which Cochrane was First Admiral of the Chilean Navy. Earle returned to Rio de Janeiro on the frigate *Hyperion* arriving on 10 December 1820, where he remained for the next three years.

Topographical drawing to aid navigation was encouraged in specialist schools towards the end of the seventeenth century. In 1673 Christ's Hospital received its second Royal Charter (Edward VI conferred the first one in 1553) to create the Royal Mathematical School whose original purpose was to train mathematicians and navigators for the navy and mercantile service.

Thomas Phillips (d. 1693), the military engineer and draughtsman, is arguably the earliest recorded British draughtsman to have produced coastal profiles. He worked for Charles II and produced designs of sea-battles and depictions of maritime fortifications that also incorporated coastal profiles and landscape views. In 'A Prospect of St. Peters Port and Castle Cornett in the Island of Guernsey, 1680' he demonstrates considerable drawing skill, and a profound knowledge of seafaring, as the shipping and craft are portrayed in convincing detail.

The eighteenth century ushered in a new systemized and regulated surveying regime. Naval officers and seamen started to produce coastal profiles and views on a regular basis in significant numbers, and some of remarkable merit. Among them Peircy Brett (1709–1781) stands out as one of the most accomplished exponents. He was also one of the earliest to effectively combine his images with descriptive text relating to his visual observations of headlands, profiles and views, and details required for safe sailing and landing.

At the outbreak of war with Spain and France in 1740 Brett accompanied Commodore George Anson (1697–1762), and became his second lieutenant, on a voyage to the Pacific with instructions to attack Spanish possessions and ships. Anson commanded the *Centurion* and by their return in 1744 they had circumnavigated the world. Among Brett's pictorial work produced on this extraordinary voyage 'A View of Streight Maire between Terra del Fuego, and Staten Island', produced in the early 1740s, is one of several examples revealing his attention to detail that would have been of significant use to the seafarer. This profile relates to a part of the world where *Beagle* would spend a large part of her time undertaking survey work.

Brett's pictures were produced for a very practical purpose and they often included extensive notes relating to location, depths, currents and wind direction, as well as other points of interest. Some of Brett's drawings were created specifically to be engraved for the official published account of the voyage. As part of his naval training FitzRoy would have been familiar with Brett's work.

Naval schools and colleges were established during the eighteenth century. Part of their curriculum was dedicated to teaching the cadets and trainee officers how to draw. They offered art tuition and art competitions to raise the standard of drawing for surveying.

In 1715 the Royal Hospital School was established at Greenwich to educate the sons of seamen. They were,

> ...instructed in Writing, Arithmetic and Navigations by a School Master appointed for that purpose; who also instructs those in Drawing who shew genius for it.... All the Boys attend the Directors, once a year to be viewed, when they bring specimens of their several performances; and three of them who produce the best Drawings after nature, done by themselves, are allowed the following premiums [prizes], according to their respective merit: First prize a Hadley's quadrant, Second prize – a case of mathematical instruments and for the Third prize a copy of Robertson's Treatise on Navigation.

During the early decades of the nineteenth century professional maritime artists were appointed to teach naval cadets in order to raise the standard of their work. John Thomas Serres (1759–1825) taught drawing at the Chelsea Naval School. His *Liber Nauticus, and Instructor in the Art of Marine Drawings* was published by Edward Orme of Bond Street, in two parts in 1805 and 1806. This was a collaborative work produced with his father, Dominic Serres (1722–1793), and was intended to aid the student of marine art (including trainee and commissioned naval officers) in acquiring the technical competence to produce authentic marine pictures.

Dominic Serres was a French emigré, born in Gascony, who settled in England. He was the only marine painter to be a founder member of the Royal Academy of Arts in London. He was well respected; in fact, so much so that he was appointed marine painter to George III. John Thomas was trained by his father and his first teaching post was at the short-lived Maritime School in Paradise Row in Chelsea, 1779–87. On the death of his father he succeeded him as marine painter to the King and also to the Duke of Clarence, the future William IV. He was also appointed as the official draughtsman to the Admiralty and in this capacity he was often away, for up to six months at a time, engaged in survey work afloat. He produced 'drawings in the form of elevations [profiles]' of the headlands and landfall of the coasts of France and Spain. Some of these were published in *The Little Sea Torch* in 1801, which contained text translated from a French sailing guide.

FitzRoy himself benefited from the tuition of a professional marine painter. In February 1818, aged 12, FitzRoy entered the Royal Naval College at Portsmouth where he excelled. As part of his naval training he received instruction from the accomplished maritime artist John Christian Schetky

Valley of Charles Island

Above: Charles Island, Galapagos by Philip Gidley King. This loosely constructed preparatory wash drawing captured the essence of the place. Under FitzRoy's editorial eye the outline and composition have been tightened up for inclusion as an engraved illustration in the *Narrative* (1839).

Opposite: Charles Island, Galapagos, engraving after Philip Gidley King from the *Narrative* (1839). The islands of Chatham and Albemarle are also featured on the same plate.

(1778–1874), who had been appointed as Professor of Drawing in 1811, a position he held for twenty-five years until the closure of the college in 1836.

Schetky's daughter, Miss S. F. Ludomilla Schetky, wrote a lively biography of her father entitled *Ninety years of Work and Play: Sketches from the Public and Private Career of John Christian Schetky* (1877). The book reveals that he was born in Edinburgh. From an early age he had a passion to join the Royal Navy, and he served as a midshipman for two years in HMS *Hind* (the same ship FitzRoy later sailed on). However, his naval service was cut short because his parents disapproved of his career choice, and so he 'consoled himself by painting the great swan-like vessels in which he longed to sail'. He was taught to draw by the Scottish landscape painter Alexander Nasmyth (1758–1840).

From the age of 14, Schetky assisted his mother in teaching drawing and painting to a class of ladies. In 1808 he taught drawing at the Military College of Sandhurst, based at Great Marlow, but after three years he applied for the position at Portsmouth 'having all my life cherished an affectionate regard for the navy'.

Schekty was a popular professor with his naval students, many of whom went on to become captains and admirals, and was also well regarded in the mainstream art community. He was known for his carefully crafted marine

(Drivestone)   CHARLES ISLAND.

(Watering Place)   CHATHAM ISLAND.

WATERING PLACE.

ALBEMARLE ISLAND.

subjects, especially large-scale sea-battles. The artist had a preference for using sepia, and many of his paintings have a predominant coppery-brown hue. He exhibited at the short-lived Associated Artists in Water-Colours; also at the Society of Painters in Water-Colours; and became a regular exhibiter at the Royal Academy of Arts in London.

He provided sketches of naval ships for J. M. W. Turner's vast oil painting of the Battle of Trafalgar, 21 October 1805. Turner (1775–1851) wrote to him asking,

> *If you will make me a sketch of the Victory…three-quarter bow on starboard side, or opposite the bow port, you will much oblige; and if you have a sketch of the Neptune, Captain Fremantle's ship, or know any particulars of Santissima Trinidada, or Redoutable, any communication I will thank you much for.*

Turner had been commissioned by George IV to paint the picture in 1822 to hang in St James's Palace. It took about eighteen months to complete. However, even with Schetky's assistance to ensure technical accuracy and detail, the picture failed to impress the King and the nautical fraternity, who wanted a more literal rather than high-Romantic interpretation of the battle. Vice-Admiral Sir Thomas Masterman Hardy, who commanded HMS *Victory* during the battle, thought the ship resembled a row of houses rather than a ship-of-the-line, as Turner had positioned her too high out of the water. The Duke of Clarence, the future 'Sailor King', took Turner to task and told him: 'I have been at sea the greater part of my life, Sir, you don't know whom you're talking to, and I'll be damned if you know what you are talking about'.

As Professor of Drawing at the Royal Naval College in Portsmouth Schetky maintained good relations with the Hydrographic Office and counted Admiral A. B. Becher, a former pupil and Assistant Hydrographer to the Admiralty during the tenures of Parry, Beaufort and John Washington, as a friend. Alexander Bridport Becher was born in 1796 and packed a great deal into his eighty years of life. He entered the Royal Naval College in April 1810. He spent most of his career at the Hydrographic Office but in his early years he participated in the surveying of the Canadian lakes, parts of the African coast, the Cape de Verde Islands, the whole of the Azores and the Orkneys.

During his time as Assistant Hydrographer he also issued the first edition of the *Nautical Magazine*, and was active in a management and editorial capacity for thirty-nine years. In its early years it was subsidized by naval and mercantile funds to about £100 per annum. The magazine was a vital reference work and included a wide range of data and observations to assist navigation. FitzRoy ordered the latest copy of it during his command of *Beagle*.

Becher recalled the influence of Schetky:

> *He brought us a new state of things altogether. We were never allowed outside the dockyard gates before he came: but he looked up the college boat directly, and got permission to take us out sketching – and such jolly expeditions as we used to have all along the coast there!*

He described his teacher as, 'A fine tall fellow he was, with all the manners and appearance of a sailor – always dressed in navy-blue, and carried his *call* [bosun's whistle], and used to pipe us to weigh anchor, and so on, like any boatswain in the service'.

Schetky was generous in his artistic advice. In 1831 he wrote to another former pupil, Captain George White, who at that time was stationed in HMS *Melville* off the coast of Portugal. He advised:

> *No doubt you are sketching every day. Now let me advise you to make minute studies of every kind of boat of the country, and draw the inside of them as well as sheer and shape – ay, the inside, with all the gear you see in them – nets, crab-pots, spars, casks, sails, and everything, most carefully, marking the colours of each, and depend upon it will find the good of it hereafter. What would not I give just now for a minute study of a Lisbon bean-code for my present picture! but I have it not and therefore I am asking everybody about their style and character.*

The naval art instructor also advised: 'Draw also, and colour, the *pescatori* and all sorts of boatmen, and always write under each sketch where they belong to, and above all, date them: it is mighty pleasant hereafter to know where you were on such a day – it is a sort of graphic log'.

Unfortunately there are no records revealing an on-going relationship between FitzRoy and Schetky. However, as an impressionable boy it is unlikely that FitzRoy would forget his mildly eccentric and enthusiastic teacher, and he certainly appears to have benefited from his art tuition.

FitzRoy's art training under Schetky may well have encouraged him to think beyond the creation of pictorial works for a practical purpose. Although in the eighteenth century Sir Joseph Banks had encouraged the inclusion of artists and draughtsmen onboard naval ships, it was not standard practice. Anyone onboard with a degree of artistic talent was engaged when necessary. But FitzRoy wanted the best results and this led him to personally pay for the artists who accompanied him on the *Beagle* voyage. They were his 'painting men'. But he was well aware of his Admiralty Instructions and the priority that dictated that craft should be placed before art.

Although many hundreds of drawings and watercolours showing coastal profiles and views, landfalls, rocks, mountains and landmarks of note, were produced throughout the course of *Beagle*'s survey work, only a small number have survived. Despite the strictures of the Admiralty Instructions, some of them do have artistic merit. They were not originally intended for display or storage in archives and museums, but for most of them that has become their fate today.

Less than thirty original coastal profiles have survived by Conrad Martens that relate to the *Beagle* voyage, and significantly less can be attributed to Augustus Earle. Part of their brief was to assist with the survey work, although both men left the voyage before it was completed. Although

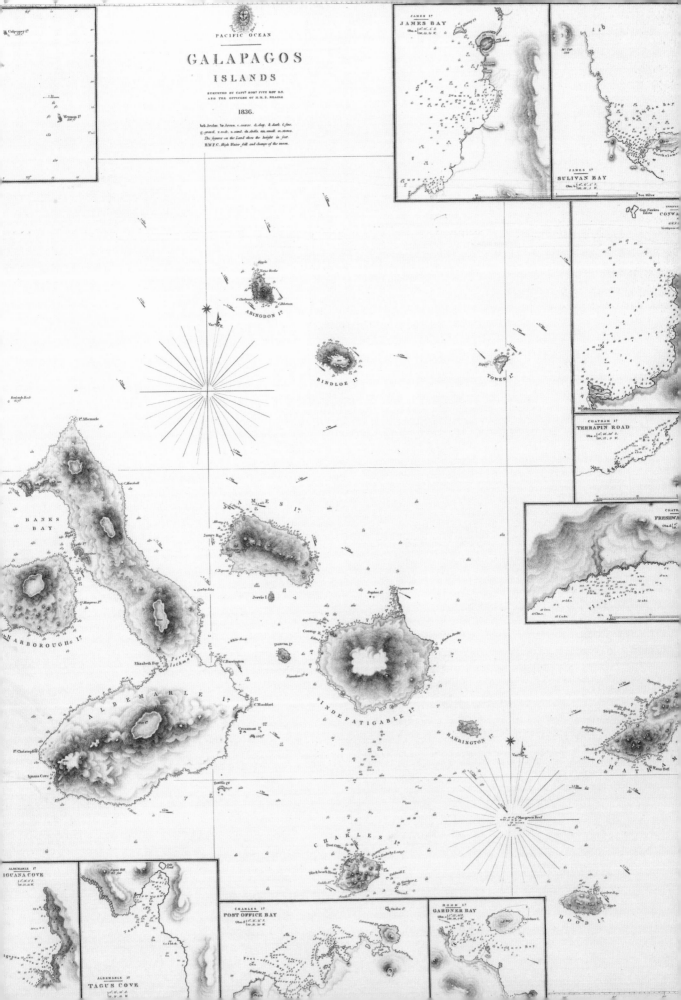

PACIFIC OCEAN

GALAPAGOS
ISLANDS

SURVEYED BY CAPT. ROBT FITZ ROY R.N.
AND THE OFFICERS OF H.M.S. BEAGLE

1836.

FitzRoy personally paid for his shipboard artists, as a formality they were approved by the Admiralty (with Beaufort's help) for the expedition; also accommodated and fed at public expense. Almost all of the pictorial work they created was deemed to be FitzRoy's personal property, while the material created by his officers that directly related to the survey and chart work was the property of the Admiralty.

FitzRoy was very impressed with John Lort Stokes's skill, commitment, and rapid work rate in producing the immensely valuable coastal profiles and views, although none of his work appears in the official *Narrative*. *Beagle*'s captain also thought highly of the work of John Clements Wickham and Philip Gidley King. He selected a series of four landscape views by King, including the islands of Charles, Chatham and Albemarle in the Galapagos, for the *Narrative*. King's view of a location where they obtained fresh water, simply named Watering Place, was also featured.

So much time, effort and expense was dedicated to producing the original profiles, views, plans and drawings for the naval charts. The costs of funding the Hydrographic Office and the survey voyages (exclusive of the Arctic and Antarctic expeditions) in 1837–8, which was placed before Parliament, revealed the sum of £68,517 – millions of pounds in todays money.

Not everyone appreciated the original works. During the mid-Victorian period the curator of the Hydrographic Office disposed of a large number of the original views. These works were a means to an end; however FitzRoy, who had laboured for so long to create (and supervise the production of) a significant number of them, would surely have disapproved.

Photography gradually took over from the role of the ship's artists and draughtsmen. The earliest photographic views were received by the Hydrographic Office in 1854, and by the early 1900s photographic records were commonplace. Even so, naval surveyors were still in service until halfway through the twentieth century.

After FitzRoy's death Admiral Sir George Henry Richards (1820–1896), Hydrographer to the Admiralty, summed up FitzRoy's contribution to naval survey work:

> *No naval officer ever did more for the practical benefit of navigation and commerce than he did… The Strait of Magellan, until then almost a sealed book, has since, mainly through his exertions, become a great highway for the commerce of the world – the path of countless ships of all nations; and the practical result to navigation of these severe and trying labours, which told deeply on the mental as well as the physical constitution of more than one engaged, is shown in the publication to the world of nearly a hundred charts bearing the names of FitzRoy and his officers, as well as the most admirably compiled directions for the guidance of the seamen which perhaps was ever written, and which has passed through five editions…*

Opposite: Admiralty chart of the Galapagos Islands (dated 1836), surveyed by Captain FitzRoy and the officers of HMS *Beagle*, bearing the stamp of the Hydrographic Office. Detailed surveys relating to specific island areas are arranged around the edges of the main chart. Depictions of coastal profiles, headlands and landfalls were arranged in such a manner on charts to aid navigation.

Chapter 6
Earle & Martens: FitzRoy's Painting Men

'Our new artist [Conrad Martens], who joined us at M. Video, is a pleasant sort of person, rather too much of the drawing-master about him: he is very unlike Earle's eccentric character. We all jog on very well together...' – Charles Darwin, writing to his sister Catherine, 20 July 1834, a hundred miles south of Valparaiso

Opposite bottom: Omai (Mai), Sir Joseph Banks and Daniel Solander, oil painting (circa 1775–6) by William Parry. Parry was a favourite pupil of Sir Joshua Reynolds, who was the first President of the Royal Academy of Arts.

Below: 'Solitude – Tristan D'Acunha, Watching the Horizon', watercolour by Augustus Earle. Earle, the self-styled wandering artist, was stranded on the remote south Atlantic island of Tristan da Cunha for eight months in 1824.

On 17 November 1831 *Beagle*'s artist was recorded in the ship's muster tables as a 'Draughtsman' under the category of 'Supernumeraries for Victuals only'. He would only receive food and drink from Admiralty funds. Augustus Earle would be the first of two professionally trained artists associated with the ship. He was a painter, panoramic artist, printmaker, writer, and poet of sorts.

Earle, however, a veteran of earlier voyages of adventure, became seriously ill (with an unidentified condition) during *Beagle*'s residence at Montevideo in Uruguay, and by August 1832 it is likely that he was no longer actively involved in the expedition. By late November 1833 his position had been taken up by Conrad Martens, who had heard of the vacancy while his ship called at Rio de Janeiro. In May 1833 Martens had sailed from England independently as a travel artist aboard HMS *Hyacinth*, commanded by Captain Francis Price Blackwood (1809–1854), although some sources suggest that the captain joined the vessel later in the voyage.

On 4 October 1833 Robert FitzRoy recorded his first meeting with Martens in an upbeat letter to Charles Darwin, who at that time was exploring the pampas near Buenos Aires: 'Mr Martens...a stone pounding artist who exclaims *in his sleep* "think of me standing upon a

pinnacle of the Andes, or sketching a Fuegian Glacier!!!" I am sure you will like him'. He further commented on Martens' manners and intelligence and compared his skills to those of Earle: 'he is a gentlemanlike well informed man. His landscapes are really good (compared to London men), though perhaps in figures he cannot equal Earle'.

Fortunately Martens' artistic career on *Beagle* lasted longer than Earle's. Even so, it was with great dismay that FitzRoy had to part company with him after being forced by the Admiralty to sell his schooner *Adventure*, the vessel he had acquired to assist his survey work. With everyone and everything (including many more of Darwin's specimens) now back onboard the *Beagle* there was insufficient living and working space. Someone had to go. Writing to Captain Beaufort from Valparaiso on 26 September 1834 FitzRoy was economical in his chosen words but the grammatical under-scoring reveals his frustration: 'My Schooner is <u>sold</u>. Our painting man Mr Martens is <u>gone</u>'.

From Valparaiso on 5 November 1834 FitzRoy wrote a letter of reference for Martens to Captain Phillip Parker King, who had settled in Sydney, which is now among the King Papers in the Mitchell Library, Sydney:

> *The bearer of this letter, Mr Conrad Martens, has parted from me, I am sorry to say, because there is no longer room for him on board the Beagle, nor money for him in his pocket. Had I more money, and more storage rooms, I should not think of ending my engagement with him. He has been nearly a year with us, and is much liked by my ship-mates and myself. He is quiet, industrious, good fellow, and I wish him well. He thinks of visiting and perhaps settling at Sydney, there-fore I write this letter by way of an introduction to you. Enclosed is a letter I received about him from Captain Blackwood of the Hyacinth. You will be able to judge of his abilities, by a glance at his works, far better than any words of mine. He has a host of views of Terrs De. [Tierra del Fuego] in his sketch book. His profession is his maintenance…*

Beagle's captain was well aware of the promotional benefits of employing artists to create original artworks. He knew that pictorial material could be translated into printed formats for a variety of publications and prints. Artistic work of the people, habits, rituals and customs, as well as the architecture, dwellings and views of exotic lands encountered during previous voyages of exploration had certainly been highly advantageous to their naval commanders, participants and patrons. FitzRoy thought of them as being of public benefit, too.

Pictorial material derived from earlier voyages had significantly raised the profiles of Sir Joseph Banks (1743–1820) and Captain James Cook (1728–1779). Banks was the first gentleman-scientist and patron to actively

Top: Conrad Martens, oil by Maurice Felton (1803–1842), painted in 1840. Felton was an Englishman who trained as a surgeon in Glasgow. He became an artist and in 1839 emigrated to Sydney where he befriended Martens. In the early 1840s *The Sydney Morning Herald* reported that 'the prominent way in which Mr Felton brings out his figures from the canvas, both faces and bust, gives them a fullness and a rotundity very opposite to the pasteboard flatness of some otherwise good artists'.

encourage draughtsmen to participate on voyages of exploration and survey expeditions. Banks paid for Alexander Buchan (d. 1769) and Sydney Parkinson (c.1745–1771), to accompany him on Cook's first voyage (1768–71) and make a visual record of all that they encountered. Both artists died during the voyage, and Banks's remarks concerning the early demise of Buchan are revealing:

> *His loss to me is irretrievable…my airy dreams of entertaining my friends in England with the scenes I am to see here have vanished. No account of the figures and dresses of the natives can be satisfactory unless illustrated by figures; had Providence spared him a month longer, what an advantage it would have been to my undertaking.*

Encouraged by the results, the British Admiralty enlisted the services of professional artists for Cook's subsequent expeditions: William Hodges during 1772–5, and John Webber in 1776–80, on the last fateful voyage, during which Cook was killed at Kealakekua Bay, Hawaii, on 14 February 1779.

Some of the pictures created during, and worked up after, Cook's voyages were exhibited at the Royal Academy of Arts in London, incorporated into published journals and accounts, and offered as individual prints for sale. This body of work confirmed and underlined the achievements of the voyages and helped to further raise the status of Cook and Banks.

FitzRoy had read these publications, seen the accompanying illustrations and admired some of the original artworks. He followed Banks's example and personally paid for his 'painting men' to accompany him on the *Beagle* voyage. However, illness and lack of space would determine *Beagle* sailed without an artist for the last two years of the voyage.

Sir Joseph Banks was the most influential civilian figure in terms of encouraging independent and British naval survey exploration from the 1760s to the early decades of the nineteenth century. More than two decades before *Beagle*'s first voyage, he was behind Captain Matthew Flinders' survey voyage to Australia in HMS *Investigator* (1801–03). As President of the Royal Society from 1778 for more than forty years he became a guide and mentor to George III, the British Admiralty and the Hydrographic Office. Banks helped secure for Charles Darwin's father a fellowship to further his medical career, because he was an admirer of Erasmus Darwin's poetical botanical verses.

Banks inherited a substantial fortune from farm lands and estates developed by his family largely in Lincolnshire. He could have gambled, womanized and drunk himself to death, but instead, he used his personality for positive ends. He became associated with, and often active within, many of the major Societies of his era. He was an eminent Member of what is now called the Royal Society of Arts (RSA). It was originally established in 1754 (without the royal prefix) by William Shipley (1714–1803), whose initial aims were to 'encourage arts, manufactures and commerce'. To pursue this goal Shipley established a drawing school in London that was attended by the Admiralty travel artist William Hodges.

The RSA offered premiums (prizes) to raise the standard of British art and design believing that it would cross over to improve manufacturing industries and British commerce. The Society of Arts organized the first London-based art exhibition of living artists in 1760, and gave a gold medal to Captain William Bligh (1754–1817) in 1793 for transporting breadfruit from the South Pacific to the West Indies. The man behind Bligh's mission was Banks. He was less interested in art for its own sake, but concerned with how it could be used to produce accurate, animated and commercially beneficial records to visually archive and promote his collections that are now housed at the Natural History Museum and the British Museum. The collections included boats, paddles and weapons, and a wide range of ceremonial, domestic and decorative items from the South Pacific islands, New Zealand and Australia, and a large collection of botanical specimens. In fact all were ultimately assembled for the benefit of Britain.

Banks attended both Harrow and Eton, and then went up to Christ Church College, Oxford. But he was not impressed with the botanical department (the professor didn't lecture) and so he hired tutors to come from Cambridge to teach him. Banks was passionate about botany, an interest he had developed as a boy. His wealth, connections and influence made him a powerful career gatekeeper. He could gain admittance for an artist, draughtsman, naturalist or scientist to a plum position on a voyage of discovery or survey expedition. The extensive Banks' letter correspondence

'Scudding before a heavy Gale off the Cape, Lat.44°', watercolour by Augustus Earle (1824). H. E. Spencer established that this view portrays the *Admiral Cockburn*, the ship which conveyed Earle and his dog Jemmy, who can be seen relaxing in the immediate left foreground, from his enforced exile on Tristan da Cunha to Van Diemen's Land (Tasmania) in November 1824.

CAPT. W.H. SMYTH, R.N. K.S.F.

(Dawson, Warren (Ed.), *The Banks Letters: A Calendar of the Manuscript Correspondence of Sir Joseph Banks...*, London, 1958) that has survived today includes many formal requests for references, employment and remuneration by men from a wide range of professions. Banks was very generous in providing assistance.

One such letter to Banks (now in The National Archives, Kew) was from Augustus Earle, who wrote from his London address at 66 Warren Street on 4 January 1818:

Sir, I take the liberty of addressing you to inform you that, I am the brother of Captain Smyth who is now employed on the survey of the Mediterranean, and on the African expedition to collect the remains of antiquity at Lepida. I accompanied him & the British Consul on their first visit to that place, and had an opportunity of making a drawing of some parts of the ruins, which H.R.H the Prince Regent did me the honor to accept of, and in a note transmitted by Sir B. Bloomfield expressed much approbation.

Earle continued in his letter plea:

I am an artist by profession and am honoured by the friendship of Captain Hurd, & hearing from him that a voyage of discovery is about to be undertaken, I am anxious for an appointment to attend it, in my profession – could I sir be favored by an interview with you, I should have an opportunity of explaining my hopes and expectations more fully, & can refer to Captain Hurd for the confirmation of what I have to say. I am young, healthy & inured to travelling by land & sea, & am ardently desirous of improvement and knowledge.

Admiral William Henry Smyth, lithograph published by Day & Haghe after William Brockedon (1787–1854). Smyth was Earle's half-brother and provided references to help him gain positions as a shipboard draughtsman.

Although Earle's pressing letter reassures Banks that he had naval connections – a reference from Captain Hurd (the second Admiralty Hydrographer) – and Royal recognition, he was on this occasion unsuccessful in obtaining a position on a voyage of discovery. The letter confirms Banks's influential position. The expedition Earle referred to was almost certainly that led by the English surgeon, explorer and naturalist, Joseph Ritchie (c.1788–1819), who early in 1818 set off with George Francis Lyon (1795–1832) to find the course of the River Niger and the location of Timbuktu. Luckily for Earle he did not go, as it turned out to be an under-funded and unproductive expedition resulting in Ritchie's death in 1819.

Despite this early setback Earle was determined to wander the world as a 'roving artist', and by the time he was signed on for FitzRoy's survey voyage in 1831, assisted by his naval connections, he had become the world's most widely travelled independent artist. Earle had briefly resided in North America, where he exhibited two watercolour portraits of gentlemen, one full length, at the Pennsylvania Academy of Fine Art. Among the many other continents and countries he visited were South America, India, the Caroline Islands, Guam, Manila, Singapore and Penang, Australia and New Zealand. He had also been shipwrecked on Tristan da Cunha. This resulted in his provocative publication, *A Narrative of A Nine Month's Residence in New Zealand in 1827: Together With a Journal of a Residence in Tristan D'Acunha, an Island Situated Between South America and the Cape of Good Hope*, in 1832. It was sparsely illustrated with engravings after Earle's watercolours.

During Earle's enforced residence on Tristan da Cunha he acted as schoolmaster to the older children on the island, and was 'unanimously appointed chaplain and read the whole of the service of the Church of England every Sunday'. He filled his time hunting and fishing, sketching and painting. His watercolours of this period rank among his finest work. They possess an emotional intensity that he rarely equals.

For two years Earle settled in Sydney, Australia. He established himself as an art teacher and set up a lithographic printing press. Earle sent back to London panoramic drawings of Sydney harbour that were used in Robert Burford's popular mechanical moving panorama located in Leicester Square. He also sent him views of the Bay of Islands, New Zealand.

Earle's half-brother, Admiral William Henry Smyth (1788–1865), would almost certainly have supplied a reference for him to secure the official position aboard the *Beagle*. Smyth had acquired an eminent reputation for his survey work in the Mediterranean. He was a friend of Captain Beaufort's. FitzRoy wanted an artist, and it made sense for him to select someone recommended by a naval superior, with an established reputation in a career-field in which he wanted to make a name for himself. When FitzRoy was

'San Salvador, Bahia', engraving after Augustus Earle from the *Narrative* (1839). Salvador (in full, São Salvador da Baía de Todos os Santos, or in literal translation, Holy Savior of All Saints' Bay) was the first colonial capital of Brazil.

SAN SALVADOR, BAHIA.

'Mole Palace and Cathedral, Rio de Janeiro', engraving after Augustus Earle from the *Narrative* (1839). The hustle and bustle of port activity is vividly captured in this print. Fortunately for Darwin, Earle had first-hand knowledge of Rio de Janeiro having been resident there some years before.

elected a Fellow of the Royal Society in 1851, among the many men who proposed him 'From Personal Knowledge' were Beaufort and Smyth.

Smyth was a noted hydrographer and astronomer who became Vice-President of the Royal Society. In the early 1830s he established an Observatory at Bedford, and is also remembered today for his *Sailors's Word-Book: An Alphabetical Digest of Nautical Terms*, published posthumously in 1867. It became a classic guide to the language of the sea and is still published today. Smyth passed down the family line a sense of adventure, discipline and achievement: his grandson was Lord Baden-Powell, the founder of the international Scouting movement.

Earle lacked the self-discipline of his brother. But he has left behind a black-and-white vision of life relating to the *Beagle* voyage, and the people and places they visited before his premature departure in the summer of 1832. Almost all of his original watercolours relating to the survey voyage are missing, and so his work is largely known through the uncoloured graphic illustrations in the volumes of the official *Narrative*. Out of what must have been a fairly large body of work only seven of his original watercolours (or perhaps drawings) were selected for publication as mixed-method steel engravings. The whereabouts today of those seven preparatory works is unknown.

Two of the printmakers, Thomas Landseer (c.1793–1880) and Thomas Abel Prior (1809–1886), selected and approved by the publisher Henry Colburn and Robert FitzRoy, were well known in their field. The London-born artist/printmaker, Landseer, spent much of his career reproducing the paintings of his famous brother, Sir Edwin Landseer (1802–1873) the animal artist, but he also established himself as one of the most gifted and innovative engravers of his generation. Prior was an accomplished printmaker days teaching students in Calais. There are no records of FitzRoy's dealings, correspondence and meetings with these men, nor with two other printmakers, S. Bull and T. Hair, who were also used by FitzRoy.

Although devoid of colour, Earle's illustrations are revealing and, in many examples, full of detail. Earle's light-hearted portrayal of 'Crossing the Line' is curiously the only recorded shipboard image of the voyage. In fact there are remarkably few depictions of life afloat by the so-called fine artists of the period. Perhaps the subjects were not thought worthy of painting, and so it has been left largely to the caricaturists such as Thomas Rowlandson, the Cruikshank family and Robert Seymour to provide images of life above and below decks. In the social pecking order of art, the caricaturists were regarded as being well below the status of the fine artists.

Earle's skills as a panoramic artist can be discerned in his *Narrative* illustration of the sweeping landscape harbour view 'San Salvador, Bahia' on page 147 (now known as Salvador). *Beagle* had arrived there in late February 1832. Everyone onboard was impressed by the magnificence of the town and this is captured in his image.

NATIVE OF KING GEORGE SOUND   NEW ZEALANDER

NEW ZEALANDERS.

'New Zealanders', engravings after Augustus Earle and Robert FitzRoy from the *Narrative* (1839). FitzRoy wanted to ensure that his images were as authentic and life-like as possible and he sought Earle's assistance to this end.

Earle conveyed the grandeur and impressive scale of the 'Corcovado Mountain, Rio de Janiero' (see page 98). In many other images he shows his talents for combining lively figures going about their daily business within architectural and landscape settings. This is evident in 'Monte Video, Mole' (see page 1), 'Monte Video – Custom House' (see page 104), and 'Mole Palace and Cathedral, Rio de Janeiro' (see page 148).

Darwin recorded his arrival at Rio de Janeiro in his Diary on 5 April 1832 of what would be an extended stay:

In the morning I landed with Earl at the Palace steps; we then wandered through streets, admiring their gay & crowded appearance

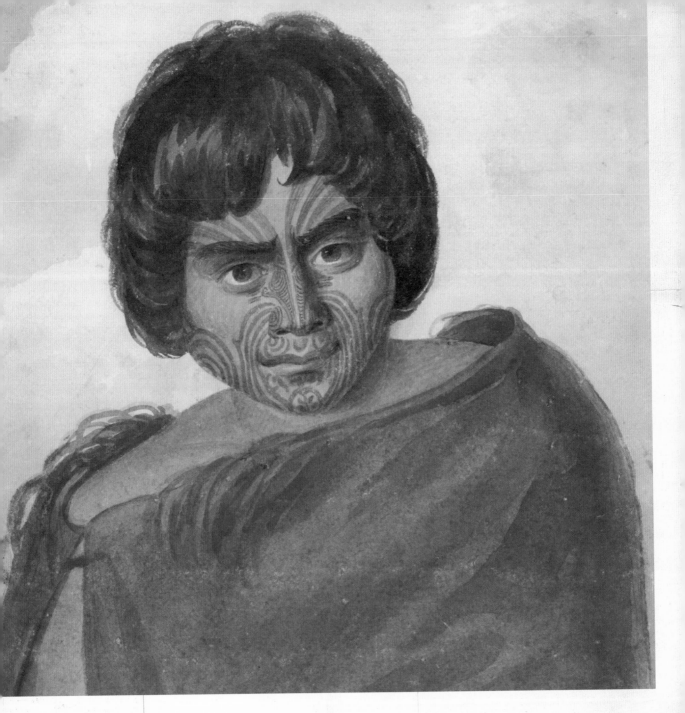

'A New Zealander', watercolour by Augustus Earle. Painted in 1827–8, this image of a celebrated tattoo artist was engraved as a frontispiece to the artist's 1832 edition of the *Narrative of Nine Months' Residence in New Zealand*… Earle recounted that 'a highly-finished face of a chief from the hands of this artist, is as greatly prized in New Zealand as a head from the hands of Sir Thomas Lawrence is amongst us [in Britain]'.

The plan of the town is very regular, the lines, like those of Edinburgh, running parallel, & others crossing them at right angles. The principal streets leading from the squares are straight & broad; from the gay colours of the houses, ornamented by balconys, from the numerous Churches & Convents & from the numbers hurrying along the streets the city has an appearance which bespeaks the commercial capital of Southern America. The morning has been for me very fertile in plans, – & at Botofogo Earl & myself found a most delightful house which will afford us most excellent lodgings.

Darwin continued:

I look forward with the greatest pleasure to spending a few weeks in this most quiet & most beautiful spot. What can be imagined more delightful than to watch Nature in its grandest form in the regions of the Tropics? We returned to Rio in great spirits & dined at a Table d'Hote, where we met several English officers serving under the Brazilian colours. Earl makes an excellent guide, as he formerly lived some years in the neighbourhood: it is calamitous how short & uncertain life is in these countries: to Earl's enquiries about the number of young men whom he left in health & prosperity, the most frequent answer is 'He is dead & gone'. The deaths are generally to be attributed to drinking; few seem able to resist the temptation, when exhausted by business in this hot climate, of strongly exciting themselves by drinking spirits.

Although Earle and Darwin became well acquainted during the *Beagle* voyage, Darwin was a little wary of the loose-living Bohemian artist. Earle was clearly a liberal minded fellow who was sexually adventurous. Darwin described him as being 'openly licentious'.

The artist's ill health is recorded by Darwin in his Diary on 9 May 1832: 'Earl, who is unwell & suffers agonies from the Rheumatism'. Writing from Montevideo on 15 August 1832 it is the Captain's turn to comment on his draughtsman's well-being as well as his artistic contribution: 'Mr Earle has been tormented by Illness, but he is now recovering rapidly and regaining his Strength and Spirits. His forte is the Human Figure, and in *that* he excels'.

For the *Narrative* illustration 'New Zealanders' (an image FitzRoy acquired from Earle after he left the voyage), the engraver Landseer worked up the image for the publication under FitzRoy's supervision and approval. FitzRoy made some astute observations on how previous printmakers and artists had dealt with this type of subject:

Few engravings, or paintings, show the real expression, features, or even colour of the Polynesian tribes. They give us half naked, perhaps tattowed man or woman; but the countenance almost always proves the European habits of the artist. The features have a European cast, quite a difference from the original, and the colouring is generally unlike; especially in coloured engravings.

FitzRoy went on to describe their appearance: 'The general complexion of both women and men is a dark, coppery-brown but it varies from the lightest hue of copper to a rich mahogany or chocolate, and in some cases almost to black. The natural colour of the skin is much altered by paint, dirt, and exposure'. FitzRoy had personally selected Earle's illustration of 'New Zealanders', and so presumably he thought it was fairly faithful. It certainly demonstrates Earle's ability as a portraitist. Above Earle's image (on the same illustration plate) FitzRoy included his own profile portraits of a New Zealander and a native of King George Sound.

More than 160 of Earle's watercolours painted throughout his career have survived, although they mostly pre-date the *Beagle* voyage. From these works we are able to gain an insight into the artist's working methods, painterly style and use of colour. The collection was acquired by Captain Smyth after his brother's death. They later became the property of the Australian-born, London-based art dealer, Rex Nan Kivell, and are now part of the National Library of Australia in Canberra.

Earle (originally spelt Earl) was born in London on 1 June 1793. His father, James Earle, was an unexceptional American painter by birth; however, his uncle, Ralph Earl(e), was a well-known eighteenth-century artist. James had left his homeland to settle in Britain as a result of his wealthy Tory family's support of Britain during the American War of Independence (1776–82).

Accounts of Earle's early life and training are sparse but he was clearly a precocious draughtsman, and exhibited at the Royal Academy of Arts in 1806. His early works were historical and classical subjects; however, he later exhibited maritime subjects relating to the Napoleonic Wars, perhaps due to the influence of his half-brother Smyth. This was a period when Benjamin West (1738–1820), the American-born, second President of the Royal Academy (succeeding Sir Joshua Reynolds), was active in his support of young men from his homeland.

West was known for his historical subjects. Among his American students (Earle's acquaintances) were the creative and inventive Samuel Finley Breese Morse (1791–1872) and Charles Robert Leslie (1794–1859). Morse became an accomplished portraitist and painter of historic scenes. He created the single wire telegraph system, and was a co-inventor of the Morse Code. Morse would have introduced Earle to life-long friend and art teacher Washington Allston (1779–1843). This influential prize-winning American painter and poet was for a time resident in London before he returned to North America in 1818. Earle's bold compositions and atmospheric colour, often combined with a high viewpoint, may derive from the influence of this artist. Leslie was born in London to American parents. He became a teacher of drawing at the West Point Military Academy, although he is best remembered as the biographer of one of Britain's best-loved landscape artists, John Constable (1843).

An earlier American student, William Dunlap (1766–1839), who had also studied with West in 1784, and became a producer, playwright, actor and historian, provides details of Earle's early life. Although not of Earle's immediate circle, Dunlap's diaries and reminiscences describe him as an 'intimate friend and fellow student'. So it is likely that Earle also studied under West: his studio was not far from Earle's childhood home at Newman Street, Marylebone.

Two other influences are clear in Earle's work. His sister Phoebe (b. 1790) married the painter Denis Dighton (1792–1827) and she also knew the Daniell family, famed as topographical artists. Dighton was a painter who occasionally followed his father, Robert Dighton (1751–1814), as a caricaturist. But he established himself as a painter of land and sea battles; one celebrated example shows the fall of Nelson on the deck of HMS *Victory* at Trafalgar (now at the National Maritime Museum, Greenwich). The exaggeration of scale and perspective apparent in this renowned work are also common stylistic features evident in Earle's work. And although Earle was not strictly speaking a caricaturist in the tradition of Gillray, Rowlandson and the Cruikshanks, his artistic liberties, which include a quirky line, exaggeration of facial features (albeit mild-mannered) and bodily gestures, reveal an affinity with that genre. Dighton copied several of Earl's pictures and produced lithographs after his work. Examples can be viewed in the British Museum, London.

William and Thomas Daniell were for a time resident at Fitzroy Square, close to Earle's home. William Daniell (1769–1837) made designs for his own publication *A Voyage Round Great Britain*, published in parts between 1814 and 1826. He also collaborated with his uncle, Thomas (1749–1840) on *A Picturesque Voyage to India by way of China* (1810). Their published works became popular and Earle may well have received some tuition from them. As Jocelyn Hackforth-Jones suggests in her publication *Augustus Earle Travel Artist* (1980): 'The Daniells' firmly constructed and clearly defined compositions, and their emphasis on accuracy, may well have influenced Earle's compositions'.

Earle also owes a stylistic debt to the flat wash drawings and watercolour patterns, often with forceful outlines, created by the British landscape and architectural watercolourists, John White Abbott (1763–1851), Francis Towne (c.1740–1816) and William Pars (1742–1782). Abbott had been a pupil of Towne. Earle would certainly have seen Abbot's work at the Royal Academy of Arts. Prints after these artists' subjects were also readily available.

Earle's collection of watercolours includes coastal profiles of Lima, Staten Island and the Island of Juan Fernandez and interior views onboard naval vessels. There are vivid portrayals of carnival time at Rio de Janeiro; the punishing of negroes; Brazilian bananas and a bat.

He spent about two years in New South Wales and there he portrayed cabbage trees, rocks, waterfalls, natives and views of New South Wales, including Government House and part of the town of Sydney. Earle painted a full-length portrait of the governor, Major-General Sir Thomas Brisbane, that was described as a 'genuine colonial production of the first class'. Other watercolours portray New Zealand chiefs, warriors, dancers, dwellings and village views, as well as images of Indian methods of transport using bullocks, or cows, in Malacca, and a catamaran in Madras Roads.

Among Earle's various views relating to his enforced stay on Tristan da Cunha is a self-portrait entitled 'Solitude – Tristan D'Acunha, Watching the Horizon' (1824), which shows him reclining on a rock, searching the sea for a rescue vessel, with his gun and dog, Jemmy, by his side. The artist's sombre and predominately grey palette heightens the drama of the scene. Earle appears neither gloomy, nor forlorn, but resigned to his fate, although his published account tells quite a different story: 'I station myself upon the rocks, straining my eyes with looking along the horizon in search of a sail, often fancying the form of one where nothing is…and again I retire to my lodging with increased melancholy and disappointment!'.

FitzRoy had high hopes for Earle, and he was clearly disappointed at the artist's early departure from the voyage. Writing in a private letter to Beaufort from Montevideo on 5 December 1833, he states: 'I do very much care for the excellent companions of my wanderings. All are well – all do well – excepting Mr. Earle who is gone from us – invalided'.

After being invalided out of the *Beagle* Earle returned to England. He made a brief sketching trip to Northern Ireland, and exhibited only three more pictures at the Royal Academy. David Stanbury (1933–1997), a passionate Darwin historian with an interest in the shipboard artists, established that Robert McCormick kept in touch with Earle, visited his London home and commissioned a painting from him. FitzRoy also corresponded with Earle and provided references for him to gain work to ease his financial woes.

On 10 December 1838 he died alone at Diana Place in London: the cause of his death was given as 'Asthma and Debility'. His replacement on board the ship would turn out to be an altogether different sort of character. In addition to his sketches and drawings in graphite, Martens also produced portraits, landscapes and marine views in colour that have survived.

Darwin's letter to Caroline of 13 November 1833 from Montevideo reveals that,

> *Poor Earl has never been well since leaving England & now his health is so entirely broken that he leaves us – & Mr Marten, a pupil of C [Copley] Fielding & excellent landscape drawer, has joined us. He is a pleasant person & like all birds of that class, full up to the mouth with enthusiasm.*

It is not entirely clear what Darwin meant by that last comment but perhaps Martens initially came across as an over-confident person. It could also mean that he was fairly restrained and generally kept his thoughts to himself. In the summer of 1834 Darwin wrote to another sister, Susan, shortly before the ship arrived at Valparaiso harbour on 23 July: 'Our new artist…is a pleasant sort of fellow, rather too much of the drawing-master about him: he is very unlike Earle's eccentric character. We all jog along very well together'.

Conrad Martens was born on 21 March 1801 at 23 Crutched Friars in the City of London. His forebears included merchants and bankers from Hamburg. His father had settled in England after serving as the Austrian consul. Martens' early education is unknown but clearly was proficient judging by his journal, letters and other writings. Two of his brothers were also artists: Henry, the eldest, with whom he maintained a regular correspondence; and his second brother John William, a topographical artist, with whom he was not closely associated.

Aged 16 Martens was taught by the extravagantly named landscape watercolourist and popular art teacher Anthony Vandyke Copley Fielding (1787–1855) at his studio in Newman Street, London. Martens' work is remarkably similar to his early teacher, and although he never loses his feeling for the picturesque, due to his involvement with the *Beagle* he

became increasingly focused on factual topography. Martens has been described as Australia's answer to J. M. W. Turner. Martens believed that there was 'no higher authority in landscape' than this British artist. He shared Turner's passion for the paintings of seventeenth-century French-born artist Claude Lorrain (*c*.1600–1682), who specialized in grand theatrical compositions with dramatic lighting effects.

Martens' tuition by Fielding was undoubtedly a turning point in his work and career. Fielding was born into an artist family and studied himself under the renowned watercolourist John Varley (1778–1842). He was a fashionable drawing master, a popular exhibitor and president of the Old Watercolour Society. The artist and art critic John Ruskin (1819–1900) was also one of his pupils and later wrote of Fielding: 'it is impossible to pass by his down scenes and moorland showers, of some years ago, in which he produced some of the most perfect and faultless passages of mist and rain cloud which art has ever seen'.

After his father's death, Martens moved to Devon, and not surprisingly many of his watercolours were of views in, and around, this county. He also sketched further afield in Kent, and he and his brother Henry exhibited at the Royal Society of British Artists at the Suffolk Street Gallery, London.

'Bay of Valparaiso looking towards Vina del Mar', watercolour by Conrad Martens. This work is not typical of Martens in terms of the portrayal of people and precise nautical details. As Richard Darwin Keynes has noted, it more closely relates to the work of Johann Moritz Rugendas, who was present with Martens during his residence at Valparaiso. This watercolour may have been given to Martens by Rugendas as part of a picture swap.

But Martens was eager to see more of the world. He was also aware that Fielding was one of many highly accomplished watercolourists and that if he decided to remain in (or return to) England, he would be in direct competition with him. After his premature departure from the *Beagle*, Martens made a conscious decision that he would not play second fiddle to Fielding and other watercolourists such as David Cox (1783–1859), of whom he would later acknowledge as a significant influence on his work; also Francis Danby (1793–1861), who excelled at dramatic lighting effects. He dropped his initial ideas of travelling to Santiago, Chile and Canton, and sailed instead for Tahiti where he stayed for seven weeks before travelling, via New Zealand, to Australia, where he lived out the remainder of his days.

It transpired to be a prudent decision because Martens carved out a successful niche as a prize-winning painter, teacher and Parliamentary Librarian. He is acknowledged as one of the founding fathers of colonial art in Australia. Occasionally he sent pictures for exhibitions in Paris and London, but most of his works were exhibited in his adopted land at several venues, including the inaugural exhibition of the Society for the Promotion of the Fine Arts in Australia at Sydney in 1847, and the Victorian Fine Arts Society in Melbourne.

In Australia Martens kept a journal of his works (now in the Mitchell Library) detailing the clients who had commissioned them, the prices paid and when the work was despatched. He used the finest materials that he could readily obtain and imported paper from the British manufacturer Whatman, and his drawing books were obtained by Ackermann and

Opposite: 'Island of Chiloe', watercolour by Conrad Martens. After Tierra del Fuego, Chiloe is the second largest island in South America. It measures 112 miles from north to south and has mountains across its entire length, creating radically different environments.

Below: 'Berkeley Sound', watercolour by Conrad Martens. Berkeley Sound is an inlet in the north east of East Falkland in the Falkland Islands. It was the preferred anchorage of the *Beagle*.

Roberson & Miller from London. From the mid-1840s he no longer had to mix his own oil paints as he used the new Winsor & Newton ready-prepared paints in metal tubes. In terms of framing his work Martens had strong views, and writing to a friend in 1854 stressed that, 'It is a downright injustice to shew a picture on which a painter's credit is staked without a frame'.

Martens' self-portrait in graphite drawn on board the *Beagle* in June 1834 reveals a bearded, studious adventurer. Darwin took time to get used to him but they became good friends. The artist was certainly industrious and by far the larger number of his illustrations – twenty-nine (signed and initialled) – featured in the official *Narrative*. They included powerful portraits of Patagonians and Fuegian families, views of natives in the Beagle Channel, as well as a depiction of the *Beagle* ashore at the Rio Santa Cruz (see page 109).

Martens effectively captured the bleakness, desolation and wilderness of many of the places visited, especially in his *Narrative* engravings of 'Mount Sarmiento, from Warp Bay', the 'Santa Cruz River, and Distant View of the Andes' (see page 108), and the 'Cordillera of the Andes as seen from Mystery Plain, near the Santa Cruz'. In these images and others, such as 'Basalt Glen – River Santa Cruz', he reveals an affinity to the Romantic notions of awe and terror popularized by the Irish-born philosopher and landscape theorist, Edmund Burke (1729–1797).

Burke popularized ideas of the beautiful and the sublime, which caught the imagination of artists, musicians and writers well into the nineteenth century in Western Europe. For Burke the 'sublime' is what has the power to compel and destroy us. The topography of large parts of South America reflected through the eyes of Martens fits perfectly with these ideological notions. Martens believed that landscape painting should be concerned 'not in that of imitating individual objects, but the art of imitating an effect which nature has produced with means far beyond anything we have at our command'. His focus on 'imitating an effect' linked him to the aesthetic ideas expounded by the eighteenth-century artist, writer and clergyman William Gilpin (1724–1804), whose book *Observations of the River Wye, and Several Parts of South Wales, etc Relative Chiefly to Picturesque Beauty; made in the Summer of the Year 1770*, had a profound impact on leading British artists.

In Martens' quest to see and paint more of the world his friend, Captain Francis Price Blackwood, had transported him to South America aboard the *Hyacinth*.

He compiled a journal of this voyage, although parts were later written up from his diary notes, which is now in the State Library of New South Wales. He continued the journal after his transfer to the *Beagle*. On 4 July 1833, the *Hyacinth* arrived in the harbour of Rio de Janeiro. Martens recorded his chance encounter with one of *Beagle*'s crew, who told him that there was a vacancy for an artist:

Little did I think at that time that I should leave Rio on any other vessel but Hyacinth or steer any other course but that for the East Indies…my conversation with Mr. Hamond [Robert Hamond], the Beagle's Mate, who was on his way back to England, and the determination to leave the Hyacinth makes me less at leisure to follow my favourite occupation [painting], my whole aim being now to find a vessel which will convey me as quick as possible from this place to Monte Video least I should by any chance miss the Beagle.

Above: 'Monte Video Harbour', watercolour and graphite by Conrad Martens. Martens made a panoramic drawing heightened with wash, dated 4 December 1833 (now in the Cambridge University Library), that he used to work up this more detailed and atmospheric watercolour.

Opposite: Caryophyllia, graphite with water-colour, found on Elizabeth Island by Conrad Martens. This is a rare example of a botanical study produced on the voyage. FitzRoy's artists focused mainly on people and places.

When Martens actually joined FitzRoy's survey ship he was sparing in his words and enthusiasm, writing on 25 November 1833: 'This day removed bag and baggage on board the *Beagle* and fairly took possession of my cabin. The weather at this time extremely hot, the thermometer as high as 90 in the shade'. The artist's journal reveals a preoccupation with the temperature of places. But the real value of his literary record is in the descriptions of locations through the eyes of an artist.

On 19 June 1834, towards the end of his time on *Beagle*, he described his first sight of Mount Sarmiento in Tierra del Fuego:

The local colour of the mountains here is generally composed of a redish brown, purple, and slate colour. The dark purple with some washes of blue serves to mark the numerous hollows and ravines, while those other parts which receive the light are generally of a light slate of ash colour, but the whole is much subdued by the intervention of a grey atmosphere which is indeed strongly characteristic of Terra del' scenery.

Four of Martens' numbered sketchbooks compiled onboard the *Beagle* have survived. They reveal a methodical working process. Sketchbooks I and III are in the Cambridge University Library, England, and II and IV are in a private collection in Australia. Although numbered, the first sketchbook in terms of chronological order is in fact sketchbook III, which includes images relating to his passage on the *Hyacinth* from England to South America. The Cambridge sketchbooks contain about a hundred images, mostly in graphite with a small number in colour.

Valley with a small stream
running into Santa Cruz River.
the hills crowned with
Volcanic Rock.
the most northern
yet discovered.

April 26 —

Above: 'Valley near Santa Cruz River',
graphite drawing by Conrad Martens.
This drawing was used by Martens to
later work up a watercolour view, now in
a private collection.

Right: 'Basalt Glen – River Santa Cruz',
engraving after Conrad Martens. Under
FitzRoy's editorial control many of the
Narrative images were altered and enlivened.
Martens' original drawing (above) and
watercolour both lack the puma chasing
the guanaco that can be seen in this view.

Martens was predominately interested in landscape views. His sketches
reveal a fascination with headlands, harbours, hills, mountains, although
architecture, ships, boats, figures of natives and occasionally *Beagle*'s crew
and animal studies all feature. He liked sketching trees but seems to have
held little interest in botanical drawing. One notable exception is his drawing
of the *Caryophyllia* found on Elizabeth Island, in the Strait of Magellan.

The *Beagle* material was mostly produced on a small scale. Martens' watercolours are worthy of detailed observation. In several examples in the collections of the National Maritime Museum, Greenwich, (donated by a FitzRoy relative) he portrays insect-sized humans going about their survey work dominated by threatening and dramatic landscape settings. His brilliant effect of using blue-grey and deep-blue highlights, sometimes to denote water, enlivens and intensifies the mood and atmosphere of the scenes.

Evidence of FitzRoy's official approval of Marten's material is evident in the sketch 'Slinging the Monkey' (see page 70). FitzRoy's initials can be seen in the upper right-hand corner, and he also annotated the sheet: 'Note the Mainmast of the Beagle a little father aft, Miz. Mast [mizzen mast] to rake more'. This shows the extent of the captain's interest in his artist's work and also the guiding hand of a professional seaman.

From these preliminary sketches and drawings it is possible to trace Martens' working method from initial subject selection to completed watercolour composition. But he did not see these sketchbooks as an end in themselves. They were visual springboards. In his 'Lecture upon landscape painting', which was delivered after he had settled in Australia in 1856, he asserted that the sketch was only the 'first part of the business', and that the 'sketch should be slight as it is for the purpose only of giving a general idea of the subject to be painted, and the beginning of the work is not time for details. It may be in pencil only, in order to show how forms combine'. He also believed that a black-and-white sketch would be suitable for arranging light and shade and that in terms of composition, 'lastly it may be composed of all together but on too small a scale to define objects or to enter into any details'.

Several of the preparatory drawings from the sketchbooks directly relate to illustrations in the *Narrative*. When compared with the sketches and watercolours, the illustrations reveal some intriguing changes. The engravers, no doubt working under FitzRoy's instructions, often altered and enhanced the originals by bringing the subjects into closer focus, and enlivening the scenes with people, and sometimes animals.

Martens' graphite sketch of the 'Valley near Santa Cruz River' (26 April 1834) shows no sign of life. The subsequent watercolour, now in a private collection, that Martens worked up is faithful to the preparatory work. However, the scene was enlivened with a puma by the engraver Thomas Landseer and was entitled 'Basalt Glen – River Santa Cruz' for the *Narrative* (shown opposite).

Above: 'Mr Nott's Old Chapel [at Papawa], Tahiti', engraving after Philip Gidley King from the *Narrative* (1839). Henry Nott was the longest serving missionary on the island. He built a chapel at Papawa, a small cove close to Matavai, which was attended by some of *Beagle*'s officers and crew.

FitzRoy described Basalt Glen as,

> *A wild-looking ravine bounded by black lava cliffs. A stream of excellent water winds through amongst the long grass, and a kind of jungle at the bottom. Lions (pumas) shelter in it, as the recently torn remains of guanacoes showed us. Condors inhabit the cliffs. Imperfect columns of a basaltic nature give it a rocky height the semblance of an old castle. It is a scene of wild loneliness fit to be the breeding-place of lions.*

Similar changes were made to the original sketches of Montevideo (8 August 1833), and 'San Carlos, Chiloe' (5 July 1834).

Writing to Beaufort from Valparaiso on 14 August 1834 FitzRoy comments, among many other things, on the movements of his artist, but at that time he was more excited about Martens' companion painter:

> *Mr Martens is at work on shore, living with an exceedingly able man of the same profession, a German by name 'Rugendas'. Pray when you have five minutes to bestow upon beautiful prints ask at a shop for 'Voyage pittoresque au Bresil' [1827–35] par M. Rugendas, published by 'Engelmann Graf & Coindet', Newman St. They are the most faithful, the very best delineations of Tropical scenery and human beings that I have seen or can imagine it possible to produce. Rugendas has been lately in Mexico, and is now here, collecting material for another work.*

Johann Moritz Rugendas (1802–1858) was a German-born artist who travelled extensively in Mexico and South America. A noted careful observer of topography and scenery, he was undoubtedly an influence on his artist-companion.

In September 1834 Martens received the unwelcome news that he would no longer be part of the *Beagle* voyage. Despite being initially upset, he embraced the opportunity and recorded in his journal: 'now more than ever at liberty to remain here or go in whatever direction my fancy should lead me'. Darwin's letter to Caroline on 13 October 1834, simply states: 'It is necessary to leave our little painter, Martens, to wander about the world'.

Darwin's association with his 'old shipmate' did not end after he left the ship. When the *Beagle* arrived in Australia in 1836 Darwin visited Martens and purchased two of his watercolours. Writing to Susan from Sydney, 28 January 1836, Darwin admitted he had been 'extravagant & bought two water-colour sketches, one of S. Cruz river & another in T. del Fuego, 3 guineas each, from Martens, who is an established artist at this place. I would not have bought them if I could have guessed how expensive my ride to Bathurst turned out'.

The Tierra del Fuego watercolour is now known by a more expansive title, 'The Beagle, Murray Narrows in Beagle Channel'. This popular image portrays a lively Fuegian in a canoe waving farewell to the *Beagle*. The watercolour still hangs in Down House in Darwin's study. Martens'

Sketchbook IV, now in a private Australian collection, contains, according to Susanna de Vries-Evans, images of Jemmy and his wife, and a drawing of the standing figure Jemmy waving from his canoe.

On 20 January 1862, after the publication of *On the Origin of Species*, Martens wrote to Darwin from St Leonards, Sydney, to congratulate him on the 'book of the season' as he called it, although he admitted he had not read it and was reluctant 'to think I have an origin in common with toads and tadpoles'. He added, 'I hope that you will not refuse another [watercolour] which I shall have much pleasure in preparing and will send you by the next mail'. Darwin duly received a watercolour of the Brisbane River.

Earle and Martens were both remarkable wandering artists. They also had a talent for writing, although at Rio de Janeiro on 5 July 1833, before he joined the survey ship, Martens' personal comments in his journal reveal a lack of confidence in his literary abilities: 'I shall not attempt a description of the place here: I am indeed but ill qualified to describe any thing but scenery, and that I am certainly better able to do with the pencil than the pen'.

Curiously both artists appear to have avoided portraying their paymaster. FitzRoy's portrait image was later taken by several photographers. He used photographic images of himself as a carte-de-visite. A competent but uninspiring oil portrait was created in later years by the Norfolk-born portrait painter Samuel Lane (1780–1859), although it was not commissioned by FitzRoy. Philip Gidley King recalled in Sulivan's *Life and Letters*: 'It was at Bartholomew James Sulivan's instigation that a portrait was painted of our estimable commander, Robert FitzRoy, now hanging on the walls of the Painted Hall of Greenwich (Hospital that was), a photograph of which has had a place in my library'. Lane was a regular exhibitor of portraits at the Royal Academy of Arts in London. He attracted a wide range of sitters; however, the earlier profile drawings and prints of FitzRoy reveal far more of his character.

FitzRoy's own interest and talent for art is confirmed by surviving drawings and watercolours and by his illustrations in the *Narrative*. Many of Captain Phillip Parker King's artworks were engraved for the first volume. FitzRoy's images included individual portraits of the Fuegians: a family group in 'Woollya' and 'Fuegians going to trade in Zapallos with the Patagonians'. FitzRoy would have benefited from artistic advice from his personal artists, and no doubt the additional skills of the engravers to enhance his original compositions. But they are evocative, powerful and compelling images.

FitzRoy also tried his hand at landscape subjects. His panoramic watercolour view of Coquimbo in Chile, dated 25 May 1835, is a good example of his proficiency as a topographical artist. Darwin wrote in his Diary on 14 May 1835: 'We reached Coquimbo, where we stayed a few days. The town is remarkable for nothing but its extreme quietness. It is said to contain from 6000 to 8000 inhabitants'. FitzRoy's view is devoid of townsfolk. Perhaps more through accident than design he has captured a sense of silence.

Overleaf: 'The Beagle, Murray Narrows in Beagle Channel', watercolour by Conrad Martens. It has been suggested that the Fuegian waving from the canoe is Jemmy Button.

'Coquimbo', watercolour, dated 25 May 1835, by Robert FitzRoy. FitzRoy has successfully captured what Darwin described as the silence of the city. Although Richard Darwin Keynes confidently asserted that this panoramic scene was the hand of FitzRoy, very few watercolours have survived that can be fully attributed to *Beagle*'s captain. Stylistically it could easily pass as the work of Philip Gidley King.

FitzRoy's artists' work complements the literary descriptions of FitzRoy, Darwin, Covington, and those written by themselves. The artists did not complete the survey expedition, and many of their original pictures are missing (including Earle's journal). However, collectively those that have survived, whether as originals or in an altered state, provide a unique visual record, offering insights into FitzRoy's quest to return the Fuegians to Tierra del Fuego and complete the survey work started on *Beagle*'s first voyage.

Writing to his sister from Montevideo on 4 December 1833 FitzRoy was eager for his relatives to share the enjoyment of the artworks created during the expedition. 'The drawings are to be distributed as they are marked, if you please. After Rice [Fanny's spouse] and yourself have looked at them sufficiently. They are done by Mr A Earle'. In a letter of 4 April 1834, he reassured his sister: 'If Mr Earle calls at your home when he returns to England – pray receive him kindly – he is a very worthy – though very unfortunate man. He will be able to tell you more in an hour than I could write in a week'.

Nora Barlow, the grand-daughter of Charles Darwin, wrote the foreword to H. E. L. Mellersh's publication *FitzRoy of the Beagle*, published in 1968.

H.M.S. Beagle

The first few lines offer a fascinating glimpse of the collection of artworks that FitzRoy had amassed during his lifetime. Barlow wrote:

> In ONE RESPECT I am perhaps in an exceptional position to write a Foreword to a Life of Robert FitzRoy, for in 1934 I went to see his daughter, Miss Laura FitzRoy, in her London home.... I remember well the look of her crowded Victorian drawing-room, dominated by a large white marble bust of her father, – a remarkable face, sensitive, severe, fanatical; combining a strength of purpose with some weakness or uncertainty, which can be more easily seen, I think, in his earlier portrait. I asked Miss FitzRoy tentatively whether the bust had ever been photographed? And I remember how her brisk answer was unequivocal and final: 'No, I should not like the idea at all.'

Barlow continued: 'There were many pictures round the walls by Conrad Martens, official artist on board H.M.S. Beagle; and I could see that the bookcases were filled with tantalizing books which must have belonged to her father'.

Unfortunately the current location of these artworks and publications is not known.

Chapter 7
The Legacy of the *Beagle*

'The Voyage of the Beagle has been by far the most important event in my life and has determined my whole career.... I have always felt that I owe to the voyage the first real training or education of my mind. I was led to attend closely to several branches of natural history; and thus my powers of observation were improved, though they were already fairly developed.' – Charles Darwin's *Autobiography*

As soon as the survey ship arrived at Falmouth, Cornwall, Charles Darwin made a hasty departure and rushed home by carriage to Shrewsbury. FitzRoy had not quite completed the expedition. *Beagle* sailed on to Plymouth, Portsmouth and Deal to receive Admiralty officials and then continued up the English Channel and along the Thames to Greenwich, arriving there on 28 October 1836 to make the last in the series of chronometric measurements to complete the meridian chain. The captain's insistence on taking so many timekeepers had paid off, as some of the chronometers had stopped or developed faults. Around half of them were working properly. The results showed a discrepancy of only 33 seconds from the expected time. This in itself was a remarkable achievement and FitzRoy was honoured by a visit from the Astronomer Royal, George Biddell Airy (1801–1892), who went on board with his wife.

Beagle then sailed downstream to Woolwich, her place of birth, where she was paid off on 17 November 1836. Over the next few years FitzRoy would dedicate himself to writing up the official accounts of the voyage. Writing to his sister he proudly related that, 'Our voyage has been more successful than I had the right to anticipate. We have been *extremely* fortunate in *all* ways. Do not be vexed at my saying that the results of it will require two years application'. In fact it would take considerably longer for him to compile, write and edit the first two volumes of the *Narrative*. Darwin completed the third volume well ahead of schedule.

Darwin's first letter to FitzRoy from his family home was nothing less than ecstatic and supportive in tone:

I arrived here yesterday morning at breakfast-time, and, thank God, found my dear good sisters and father quite well.... Indeed all

England appears changed excepting the good old town of Shrewsbury and its inhabitants.... I do hope that all your vexations and trouble with respect to our voyage, which we now know HAS an end, have come to a close. If you do not receive much satisfaction for all the mental and bodily energy you have expended in His Majesty's Service, you will be most hardly treated.

FitzRoy's reply was equally enthusiastic:

I was delighted to see that the Valpo [Valparaiso] cargo of charts had not only arrived but they were mostly Engraved – or in the Engraver's hands – and on a large scale! They have given much satisfaction at the Hyd [Hydrographic] Office.... I was delighted by your letter, the account of your family – & the joy tipsy style of the whole letter were very pleasing. Indeed Charles Darwin, I have also been very happy – even at the horrid place Plymouth – for that horrid place contains a treasure to me which even you were ignorant of!! Now guess – and think & guess again. Believe it, or not, – the news is true – I am going to be married!!!!!! To Mary O'Brien.

Emma Darwin, watercolour by George Richmond. This portrait was painted in March 1840 as a companion picture of her spouse on page 86.

Everyone was surprised by the news of FitzRoy's impending nuptials, although Darwin had commented in a letter to his sister Caroline on 12 November 1831 before the *Beagle* had departed from England: 'if I believe all I hear the Captain is as perfect as nature can make him – it is ridiculous to see how popular he is, ladies can hardly splutter out big enough words to express their big feelings'.

Afloat, all the evidence pointed towards FitzRoy being an emotionally restrained man who would struggle at forming a relationship with a member of the fairer sex. In fact FitzRoy would marry twice. Little is known of Mary Henrietta O'Brien (1812–1852) except that she was the daughter of Edward James O'Brien, a retired senior Irish army officer who reached the rank of major general. FitzRoy is believed to have married her on 8 December 1836. After her death in 1852, he married his second wife, Maria Isabella Smyth (1819–1889), with whom he was distantly related through a cousin, on 22 April 1854, in London.

With his first wife, FitzRoy had three daughters, Emily-Unah (d. 1856), Fanny, and Katherine; and a son, Robert O'Brien FitzRoy, who became a vice-admiral and was knighted. His second wife produced a daughter, Laura Maria Elizabeth. FitzRoy's first marriage in 1836 coincides with a radical change in his religious convictions. He became a fundamentalist Christian. From this time forward he would not doubt the veracity of any of the biblical stories.

On 29 January 1839 Darwin married Emma Wedgwood (1808–1896), his first cousin, and the youngest daughter of Josiah Wedgwood II, at St Peter's

Above: Down House, Kent, watercolour by Albert Goodwin (1845–1932). Goodwin painted several views of the family home. He was influenced by the Pre-Raphaelites, and was championed by the controversial Victorian critic John Ruskin.

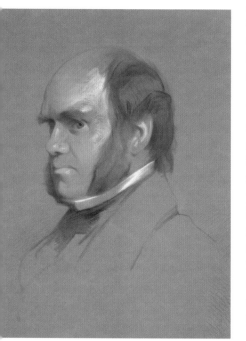

Church in Maer, Staffordshire. He weighed up the pros and cons of marriage in a list that has survived to this day. The second key point in the 'pros' column noted: '– Constant companion, (& friend in old age) who will feel interested in one, – object to be beloved and played with. – better than a dog anyhow'. Darwin was also devoted to his dogs Bob and Polly.

Emma was charming with an abundance of patience. As a child she was messy and this led to her nickname 'Little Miss Slip-Slop'. She was musical and for a time studied piano with the celebrated composer Frederic Chopin (1810–1849) in Paris during her grand tour of Europe. Emma was also adept at some outdoor sports and was an accomplished archer.

Together Emma and Darwin had ten children, seven of whom would survive into adulthood. After returning from the expedition, Darwin rented rooms in Cambridge and London to enable him and his assistant, Syms Covington, to arrange for further study and development of the notes of his various collections. But London at this time was not an ideal place for family life, or for his health, and the rapidly expanding Darwin family eventually left their small rented house in Upper Gower Street near Bloomsbury and opted for a more tranquil rural residence in the county of Kent. In 1842 they bought, with the help of family money, Down House in the village of Downe, despite Darwin's opinion that it was 'very solid throughout, though oldish and ugly'. It provided easy access to central London, only 16 miles away. Today it is open to the public for visits and is managed by English Heritage.

Throughout the *Beagle* voyage Darwin had been a fearless man of action – riding for hundreds of miles with the gauchos, sleeping rough, smoking *cigarittos*, actively participating in the arduous survey and exploration work and helping to calm potential political revolutions in South America. He experienced a period of poor health during the voyage but it passed. It remains a mystery as to the exact nature of the illness that later took hold of him. After moving to Down House he led a semi-invalid existence perhaps brought on by the psychological pressures of his work on evolutionary development. His wife was a loyal and constant nurse.

Darwin's 'sand-walk' of around a quarter of a mile, which was a pathway laid out by him beside his property, became part of his regular daily exercise regime that also offered valuable solo thinking time. He also placed his constitutional faith in the fashionable treatment of the cold baths and water spa treatments in Malvern, Moor Park in Surrey, and elsewhere. The treatments appeared to work, although their beneficial effects were short lived. Sadly they could not help his favourite child, Annie, who died at the age of 10 on 23 April 1851. She was buried in Malvern. Infant and child mortality were common during this period and FitzRoy also lost a son. Both men responded in profoundly different ways. FitzRoy hardened his religious beliefs: Darwin would from time to time walk his family to church but return home alone.

The irony of the profession that Darwin was going to pursue in his pre-*Beagle* days was not lost on him. He was initially teased by *Beagle*'s seamen for his assured conviction in the verity of the biblical stories. Writing in later years he declared that his interest in becoming a clergyman gradually waned and died a 'natural death'. Upon his return, with the help of his sisters, he prepared his father for his career change to a life of science.

Darwin's collections were creating great excitement among the scientific community. John Stevens Henslow was delighted with his pupil's success and introduced him to Sir Charles Lyell (1797–1875), the President of the

Above: Darwin's Certificate of Candidature for Election to the Royal Society in 1839. He was personally recommended by, among others, George Peacock and Sir John Herschel.

Opposite bottom: Charles Darwin, pastel drawing, early 1850s, by Samuel Laurence (1812–1884). Laurence excelled at portraits of Victorian notables. Sitters included Charles Babbage, Charles Dickens, Adam Sedgwick and William Makepeace Thackeray.

Geological Society. They became firm friends and worked closely together thereafter. The Oxford-educated Scotsman became a baron in 1864. His 'methods and style' caused a sensation in Victorian society and greatly influenced many prominent scientists besides Darwin. Professor Henslow had urged caution in the reading of Lyell's *Principles of Geology* (Volume 1). The publication suggested that the age of the earth was significantly different to biblical interpretations.

Darwin would be elected as a Member of the Royal Geological Society and many scientific organizations including the Royal Society. He would receive a congratulatory letter from his hero, Alexander von Humboldt, although he was disappointed with him when they finally met in London. Darwin also met and was assisted by Richard Owen (1804–1892), who was Hunterian

Rhea Darwinii

Professor at the Museum of the Royal College of Surgeons, and also became Director of the Natural History Museum. However, Owen did not support Darwin's radical theories. In the USA, Darwin's work was championed by the pioneering botanist Asa Gray (1810–1888).

The collections of birds, insects and mammals were presented to the Zoological Society where Thomas Bell (1792–1880) examined the reptiles, George Robert Waterhouse (1810–1888) looked at the insects and mammals and John Gould (1804–1881) the birds. Darwin's labelling of the Galapagos birds had been uncharacteristically slapdash and he asked FitzRoy to provide accurately labelled examples from his own collection. Of course FitzRoy obliged. Almost all of *Beagle*'s officers and some of the crew had formed small personal collections of sorts.

Gould was an English ornithologist and renowned artist-illustrator (although his wife Elizabeth was the better artist, helping him to draw, lithograph and colour many of the images), who was the first Curator and Preserver at the Museum of the Zoological Society. At a meeting there in January 1837 he reported that among Darwin's collection of birds from the Galapagos he had identified a group of finches. In fact they were 'a series of ground Finches which are so peculiar as to form an entirely new group, containing twelve new species' (Darwin's field notes on the Galapagos: 'A little world within itself'). The species were unique to particular islands and clearly had developed in relation to their environment, their beaks varying from strong stout ones for crushing, to thin and tapering examples for extracting grubs, sometimes with the aid of a short spike to spear the food. Gould also identified a bird which had been shot by Conrad Martens at Port Desire (it had been partly eaten but rescued by Darwin) as a new species that he named *Rhea darwinii*. It resembled an ostrich. In fact, Darwin and Gould did not realize that the French naturalist Alcide d'Orbigny (1802–1857) had previously identified the creature and named it *Pterocnemia pennata*. However, Darwin now had more pointers to develop his theory of evolution via natural selection.

On July 1837 he recorded in his private journal that he had 'opened first note Book on "transmutation of species". Had been greatly struck from about month of previous March on character of S. American fossils and, Species on Galapagos Archipelago. These facts origin (especially later) of all my views'.

Darwin was profoundly influenced by economist Thomas Robert Malthus's *Essay on the Principle of Population* (1798). Malthus's work highlighted that population growth would always exceed the food supply, creating perpetual states of hunger and disease. Darwin adapted his ideas of the natural perpetual struggle for survival into his own evolutionary theories.

Another prominent figure was the English botanist and explorer Joseph Dalton Hooker (1817–1911), who had served as the naturalist on James Clark Ross's Antarctic expedition of 1839–43 in HMS *Erebus*. Hooker visited Down House and he was the first person to whom Darwin confided his theory of Natural Selection.

Oppostite inset: Beaks of finches, wood-engraving. This illustration featured in the early editions of *The Voyage of the Beagle*. The beaks of these Galapagos finches developed differently in response to their environments on different islands.

Opposite: Rhea darwinii, coloured lithograph from *Birds* Part 3 No. 5 of *The Zoology of the Voyage of H.M.S. Beagle* by John Gould. Darwin and his shipmates were tucking into a hearty meal after shooting this creature when he suddenly realized it was a new species. What remained was shipped back to England.

Below: Caricature of Charles Darwin, from the magazine *Vanity Fair*, 30 September 1871.

Darwin was wary of what British Society and the Establishment would make of his theories. He was in no rush to publish his findings. He was also concerned about the reaction of his wife, who was a devout Christian and to whom he had written a secret letter to be opened in the event of his premature death, stating his wishes that the manuscript containing his theory of evolution by means of natural selection should be published and that a sum of up to £500 should be released from his estate to assist with the publication.

But the letter he received on 18 June 1858 from the roving naturalist and specimen collector, Alfred Russel Wallace, came as a bolt out of the blue. Wallace had developed virtually identical theories concerning evolutionary development independently, although at a later date, and these were enclosed in the letter. This spurred Darwin into action and with the help and support of close colleagues and friends he decided to develop his preliminary drafts into a publishable form that was given the title *On the Origin of Species by Means of Natural Selection, or The Preservation of Favoured Races in the Struggle for Life* (it has become universally known by the shortened title of *The Origin of Species*). This groundbreaking work was available in bookshops late in November 1859 and immediately sold out. Wallace acted with good grace and deferred to Darwin recognizing that he had been pipped to the post.

The book was published by John Murray (1808–1892), part of the distinguished Murray family publishing dynasty, who was an amateur geologist. His father had published Sir Charles Lyell's geological works and yet Murray was wary of Darwin's book and the underlying controversial argument. After seeking a second opinion he reported to the author that it would be preferable to write a book on pigeons, as there was a great deal of popular interest in this subject. Fortunately he had second thoughts: the book has endured and remains in print in various formats and almost all languages.

FitzRoy was furious at the book's publication and could not forgive himself for unwittingly aiding his messmate to collect the evidence to challenge what he (and many others) held as a core belief that God created the world. Of course, they believed science had a part to play in life but in their resolute opinion it could offer no credible explanation of Creation and evolutionary development.

As Rear-Admiral G. S. Ritchie, Hydrographer of the Navy and the author of *The Admiralty Chart* (1968), succinctly summed up:

> *Whereas Darwin saw the shells and sea-washed pebbles of the plains of Patagonia as evidence of the instability of the earth's crust, FitzRoy could visualise no 'improvement' of Man since creation and believed that there were no separate beginnings of savage races but that all were direct descendants of Adam and Eve.*

Darwin's theories suggested that man was descended from apes. This was too much for the wife of the Bishop of Worcester, who commented: 'Descended from the apes! My dear, let us hope that it is not true, but if it is, let us pray that it will not become generally known'.

The infamous Oxford Debate on Saturday, 30 June 1860, organized by the British Association for the Advancement of Science (BAAS) and chaired by John Stevens Henslow, was held at the Oxford University Museum of Natural History. Darwin disliked public speaking and claimed he was too ill to attend. But his supporters were present and ready to speak his cause. The attack against evolution was led by the eloquent Anglican Bishop Samuel Wilberforce (1805–1873), known as 'Soapy Sam' for his habit of rubbing his hands as if washing them and his slippery debating style. He had been primed by Richard Owen, who had become jealous of Darwin's ascendancy within the scientific community. He focused on one of Darwin's supporters, the biologist Thomas Henry Huxley (1825–1895). Huxley was largely self-taught and became one of the finest comparative anatomists of his era. He was assistant surgeon on HMS *Rattlesnake*'s surveying expedition to New Guinea and Australia in the 1840s under Captain Owen Stanley, the former captain of HMS *Britomart*. Huxley was the man who coined the term 'agnostic' and who described himself as Darwin's bulldog. He was well equipped to take on the Bishop.

Accounts differ as to what exactly was said. It is believed that Wilberforce asked Huxley if he was related to an ape on his grandmother's or grandfather's side, and Huxley's reply to have been: 'I would rather be the offspring of two apes than be a man and afraid to face the truth'. Huxley wrote down a different and less dramatic version:

> *If then, said I, the question is put to me would I rather have a miserable ape for a grandfather or a man highly endowed by nature and*

Robert FitzRoy surrounded by family and friends, circa 1860, by an unknown photographer.

Certificate of a Candidate for Election.

(N.B. Directions for filling up Certificates are given on the other side of this leaf.)

(Name) *Robert Fitzroy.*

(Title or Designation) *Member of the Royal Geographical Society.*

(Profession or Trade *) *Captain in the Royal Navy.*

(Usual place of Residence) *Norland Square, Bayswater,*

The Discoverer of

The Author of *Narrative of a Ten Years' Voyage of Discovery round the World, written in Connexion with Capt. P. P. King,*

The Inventor or Improver of *a Surveying Quadrant,*

Distinguished for his acquaintance with the science of *Hydrography & Nautical Astronomy,*

Eminent as a *Scientific Navigator, & for his chronometric measurements of a chain of meridian distances during the circumnavigation of the globe which he conducted.*

being desirous of admission into the **ROYAL SOCIETY OF LONDON**, we, the undersigned propose and recommend him as deserving that honour, and as likely to become a useful and valuable Member. Dated this 5 day of *June* 18 51.

From General Knowledge.

Edward Forbes
Richard Owen
C. Wheatstone.

Elected
CRW.

From Personal Knowledge.

F Beaufort
W H Smyth,
J Lloyd
Ph. J. Yorke
John Richardson
J. Badenton
J Walker
John Edw Gray
Robert Harry Inglis
Charles Darwin
J.H.G.

Read to the Society on the 27 day of February 1851
To be Balloted for on the 5 day of June 1851

* If no Profession or Trade, this should be stated by filling up the Blank with the word None.

Above: FitzRoy's Certificate of a Candidate for Election to the Royal Society in 1851. Among the people who recommended FitzRoy from personal knowledge were Captain Francis Beaufort, Admiral William Henry Smyth and Charles Darwin.

Opposite: FitzRoy's barometer, which included a thermometer and a hygrometer.

possessed of great means of influence & yet who employs these faculties & that influence for the mere purpose of introducing ridicule into a grave scientific discussion, I unhesitatingly affirm my preference for the ape.

All accounts agree that Huxley won the debate, although contrary to popular belief, he did not take on and defeat the bishop singlehandedly. Joseph Dalton Hooker deserves recognition for his active and more significant participation in the debate, countering Wilberforce's argument to great effect. Among the shouts and screams one man was barely recognizable within the packed gathering. According to some reports Robert FitzRoy was clutching a Bible. His words were inaudible but he was adamant that Darwin was wrong. He wrote to his former messmate: 'My dear old friend – I, at least, *cannot* find anything "enobling" in the thought of being a descendant of even the most ancient *Ape*'.

Since leaving the *Beagle* FitzRoy's career had been spectacularly varied, although frustrating in terms of public and professional recognition. His posts and positions were largely unfulfilling. He was spending more and more of his private family income for what he believed to be for the public benefit. FitzRoy became Member of Parliament for Durham in the north east of England in 1841, a position arranged for him by an influential relative. During his tenure he helped to lay the groundwork for the introduction of the Mercantile Marine Act of 1850 that stipulated that captains and first mates of all vessels must have certificates of competency.

During his brief political career he was appointed an Elder of Trinity House, which had responsibilities for looking after lighthouses, lightships, beacons and buoys. He was also Acting Conservator of the River Mersey. But FitzRoy gave up all of these posts for the opportunity to become the governor of New

Zealand. The motivation was not money: he would have been financially better off remaining in Britain. Yet again he was driven by a sense of duty and a genuine belief that through his efforts he could make a difference. He had been impressed with the missionaries' work in New Zealand during his command of the *Beagle*, and the missionaries had offered him moral support in justifying his actions regarding the Fuegians.

FitzRoy's family arrived in New Zealand in December 1843. Prophetically he had written to Captain Phillip Parker King stating that 'trusting in Providence – and anxious to raise the New Zealanders. I anticipate no great difficulty with them – but abundant trouble with the whites'. He tried his best to balance the interests of the Maori chiefs and the white settlers and meted out justice that he believed to be fair. According to some of the white settlers he was too lenient on the Maoris.

Governor FitzRoy was out of his natural element. The government and leadership of a shore-based establishment was not comparable to the command of one of Her Majesty's ships. In addition to the internal politics that led to the burning of Union Jack flags, complaints were made against him. The colony had also run up substantial debts. Partly as a scapegoat, FitzRoy was dismissed from the post after losing the confidence of all interested parties. On November 1845 he handed over responsibility to George Grey, a former naval captain and explorer, who earlier in that year had been made governor of the Cape Colony and High Commissioner of South Africa.

In his *Autobiography* Darwin wrote: 'I saw FitzRoy only occasionally after our return home, for I was always afraid of unintentionally offending him'. FitzRoy's last recorded visit to Down House was in 1857. However, Darwin's letter to FitzRoy from Down House, dated 1 October 1846, reveals that (at this time) he was still close to this former captain.

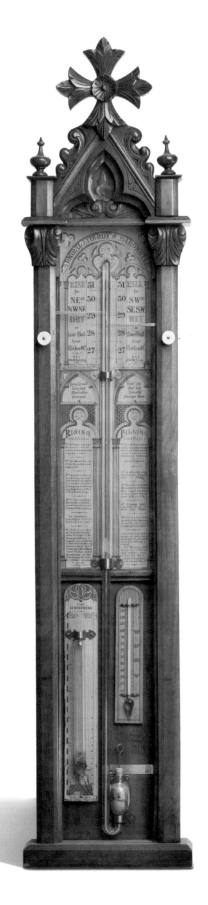

> *I did not hear for more than 4 weeks after your return [from New Zealand] that you were in England, and now, though I have nothing particular to say – I cannot resist writing to congratulate you on your safe arrival after your bad passage home, and to express my most sincere hope that Mrs. FitzRoy and your family are all well – I was in London yesterday – for the first time – since I heard of your return and found your address at the Admiralty & in Dover St. the servant told me to direct as I have done. I got your pamphlet the other day and was very much interested by it, for I had heard comparatively little about New Zealand; I fear you must have undergone much trouble & vexation and been ill repaid except by the consciousness of your own motives, for the sacrifices which I am aware you made in accepting the Governorship.*
>
> *I hope that your health has not suffered and that you are as strong & vigorous as formerly – I have hardly the assurance to ask you to spare time to write to me, but I should be very glad to hear about yourself Mrs. FitzRoy & the children. My life goes on like Clockwork, and I am fixed on the spot where I shall end it; we have four children, who & my wife are all well. My health, also, has rather improved, but I am a different man in strength and energy to what I was in old days, when*

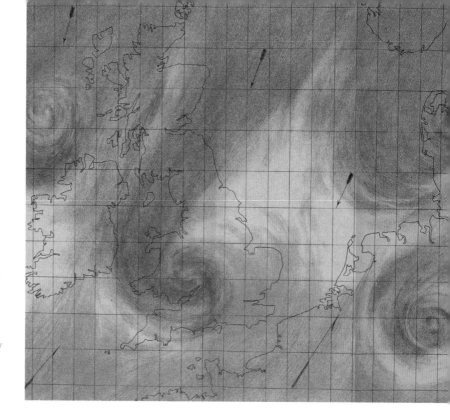

Tropical and Polar Air Currents Plate VII, from Vice-Admiral Robert FitzRoy's *The Weather Book: A Manual of Practical Meteorology* (1863). This is the earliest depiction of warm and cold air masses on a map. FitzRoy's first official storm warning was issued on 6 February 1861 and telegraphed to a number of shore stations. Warnings were sent to fifty locations in the first year, and 130 in the second. Eventually the system would be accepted and save thousands of lives.

I was your 'Fly-catcher', on board the Beagle; I have just finished the 3rd. & last part of the Geology of the Voyage of the Beagle – viz: on S. America; I will direct a copy (in a weeks time when published) to be sent to the Carlton Club, without you prefer it being sent elsewhere. – It is purely geological & dull enough, but I hope it contains some few new facts & views – I have now with the exception of some Zoological papers on the lower marine animals completed all which I shall ever attempt on the materials collected during the voyage.

I am aware how little chance there is of your having time to spare, but if ever when in Town Mrs. FitzRoy & yourself should feel inclined to spend a few days in the country – it would give my wife & myself real pleasure; – we have a tolerably comfortable house in a very quiet, retired, airy part of the country –

I think I ought to apologise for the length of this letter. – Pray give my kind & respectful regards to Mrs. FitzRoy & Believe me, dear FitzRoy

Yours truly & obliged Charles Darwin

The Admiralty clearly did not believe that FitzRoy was entirely to blame during his governorship of New Zealand and he continued to receive a series of official appointments. In September 1848 he was made Superintendent of the Royal Dockyard at Woolwich and in March of the following year was given his last seafaring command to oversee sea-trials of HMS *Arrogant*, the first British naval frigate to be powered by both sail and screw-propulsion. This type of technology had been advocated by Isambard Kingdom Brunel (1806–1859) and developed by Sir Francis Petit Smith (1808–1874). FitzRoy was proud to take his own son on board (aged 10, Robert O'Brien FitzRoy already wanted to pursue a career in the navy). Initially things went well for him and the men were happy with his command. However, the captain's depression returned when the ship arrived in Lisbon. The trials were not complete, but FitzRoy was exhausted, worried about money troubles, and this time he did relinquish command.

FitzRoy's influential connections saved him again and in 1851 he was appointed a Fellow of the Royal Society, supported by, among others, Beaufort and Darwin.

On 5 November 1852 FitzRoy wrote to Lord Dynevor (his sister's husband) proudly informing him of a 'Tidal Expedition' being arranged by the Admiralty. Beaufort had asked FitzRoy to lead it. The expedition was 'to visit many places in the Atlantic– and, eventually in the Pacific. The British Association [for the Advancement of Science] and Royal Society have asked for it'. For some reason, probably economic, the expedition never took place.

In 1854, on Beaufort's recommendation, he took up the new position relating to the collection of weather data, which had the grand title of Meteorological Statist to the Board of Trade. He invented a barometer (subsequently named after him) that was sent to various weather stations (on land and onboard ships) in and around Britain and abroad to assist with data collection, and also developed a system of warning signals for shipping. FitzRoy's *The Weather Book* was published in 1863. He would receive £200 for the sale of the book's copyright, an insignificant sum to resolve his massive expenditure for what he perceived to be of public benefit. His weather forecasts were published in *The Times* newspaper as they continue to be to this day. Then and now we all know of the variable accuracy of these forecasts and people poked fun at them. FitzRoy was unfairly blamed.

In more recent times Paul Simons, reporting for *The Times* (5 October 2007), wrote a memorable piece entitled, 'If only forecasters could speak the same language as the rest of us – The Met Office, still making heavy weather of it'. Simons observed:

> *Ever since Vice-Admiral Robert FitzRoy set up the world's first national meteorological office in England in 1854 forecasters have been blamed for getting the weather wrong.... So you have to feel sorry for Michael Fish [a BBC weatherman]. Every anniversary of the Great Storm of October 1987, out trot the repeats of his legendary forecast about 'don't worry, there is no hurricane on the way,' just hours before the big storm struck.... So there was national horror when 100mph winds felled 15 million trees, power and transport links collapsed and 19 people died, the most devastating storm in England for more than 20 years.*

Fish and FitzRoy did the best they could; the former with a sophisticated satellite system to help him. But one thing is for certain, the weather is by its very nature capricious, and by the time you are listening to or looking at the latest weather forecast it could well be on course for an alternative outlook. In the early 1860s FitzRoy's health suffered through the carping criticism, and although he was semi-retired, he was still subjecting himself to a heavy work regime.

Opposite: Base of the skull of *Toxodon Platensis* from *Fossil Mammalia* Part 1 No. 1 of *The Zoology of the Voyage of H.M.S. Beagle* by Richard Owen. The Toxodon is an enormous extinct mammal, similar to the hippopotamus. This skull was one of many spectacular fossils Darwin brought back from South America. It was discovered by boys in a remote village in Uruguay who had been using it for target practice. Darwin purchased it and by chance discovered a tooth that exactly fitted one of the sockets in the skull 200 miles from the site.

FitzRoy had shown sympathy towards Darwin during his prolific periods of seasickness, and had held back *Beagle* from sailing on one occasion when Darwin appeared seriously ill to enable him to recover on shore for a month. Darwin's letter to his former expedition captain, dated 28 October 1846, was heartfelt:

Farewell, dear FitzRoy, I often think of the many acts of kindness to me, and not seldom of the time, no doubt quite forgotten by you, when, before making Madeira, you came and arranged my hammock, with your own hands, and which, as I afterwards heard, brought tears to my father's eyes.

However, FitzRoy felt personally let down by his former *Beagle* companion when he viewed Darwin's proofs for the third volume of the *Narrative*. In the preface Darwin had failed to fully acknowledge the significant role of the captain and officers in terms of his own personal success experienced during the voyage. Darwin was normally an even-tempered gentlemen but he did not take kindly to the diplomatic rebuke from his 'beau ideal of a captain'. He included the appropriate credits to acknowledge his shipmates, but in his *Autobiography* he also recorded his irritation, blaming the captain's bad temper and unpredictable behaviour. In this regard Darwin was wrong: he owed a huge debt of gratitude to many people, not least to his old messmate.

Darwin's *Autobiography* was not originally intended for publication. He compiled a series of private reminiscences and recollections for the interest of his family and future familial generations. As it was not intended for public eyes we should excuse the haughty tone. Darwin is critical of friends and colleagues, and he comes across as an amiable though self-centred man. He acknowledged that his illness removed him from Society, but this also enabled him to become self-absorbed, providing the means to achieve a wide range of work. Visitors tired him and made him ill, although he did occasionally meet and exchange photographs with *Beagle* shipmates, including Sulivan, Mellersh and Wickham. They visited Down House on 21 October 1862. Darwin also corresponded with King and Stokes.

Darwin co-ordinated his extensive scientific enquiries from home, too, writing and receiving tens of thousands of letters. His charm was used to good effect in engaging everyone around him to help him with his work. People appeared to oblige without resentment.

Darwin was surprised and shocked when FitzRoy did not appraise him of the additional chapters he added to Volume II of the *Narrative* before publication that included 'A Very Few Remarks with Reference to the Deluge'. FitzRoy's remarks were essentially a counter-blast for the benefit of seamen to what he believed were the specious underlying hints and suggestions of Darwin's words that challenged long-standing religious beliefs. In fact, the critics scoffed and laughed at FitzRoy's old-fashioned biblical views. It now seemed to most scientifically minded people that the world was significantly older than the chronological dates suggested in the scriptures. To them no part of the evolutionary history of the world could be explained in any way by a

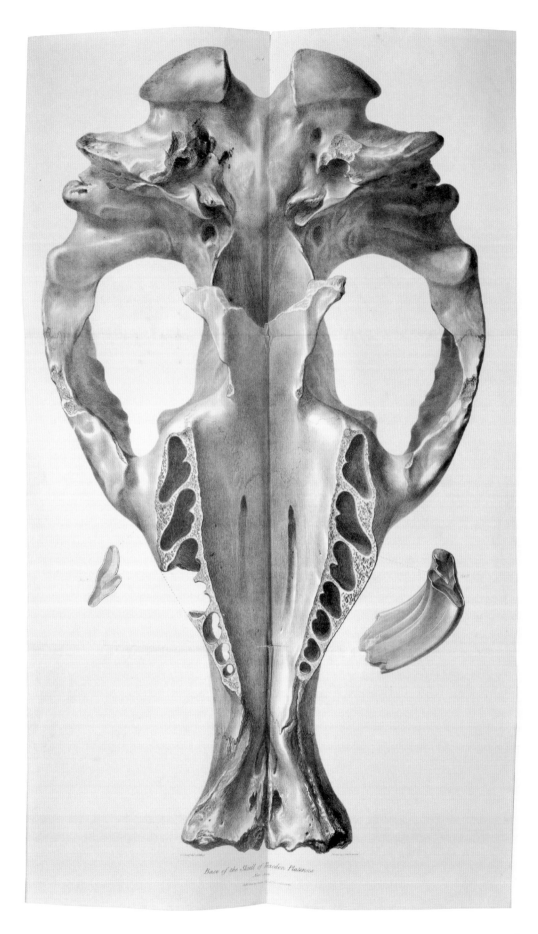

Base of the Skull of Toxodon Platensis

Canis antarticus from *Mammalia* Part 2 No. 1 of *The Zoology of the Voyage of H.M.S Beagle* by George R. Waterhouse. Darwin was sure that the Falklands fox was a unique species. FitzRoy disagreed with his view. Waterhouse was curator of the Zoological Society of London's museum, and later keeper of Mineralogy and Geology at the British Museum.

deluge or flood, other than that more dramatic forces were at work and had been for millions of years (maybe more).

In 1863, through seniority, FitzRoy was promoted to Vice-Admiral, but this made no difference to his state of mind. He descended into the deepest melancholy. Friends were concerned for his health, but they could not predict the tragic outcome. Doctors advised 'total rest, and entire absence from his office for a time'. Before breakfast on Sunday, 30 April 1865, Vice-Admiral Robert FitzRoy locked the door to his dressing room and with his razor cut his own throat.

FitzRoy had spent his entire family fortune pursuing public service. A benevolent fund was arranged by Bartholomew James Sulivan, his colleague at the Board of Trade which supervised the work of FitzRoy's 'Met Office'. Sulivan remained loyal to his former commander and sensitive to his feelings. £3,000 was raised to pay FitzRoy's family back for the monies he expended in public service. Darwin personally contributed £100. Queen Victoria granted a grace-and-favour apartment at Hampton Court Palace for FitzRoy's widow and daughter. The captain of the *Beagle*'s final resting place was near his home at All Saints with Saint Margaret Churchyard in Upper Norwood, close to a road. Partly due to FitzRoy's ill health, the family had moved from their usual London residence of 38 Onslow Square to Lyndhurst House, Upper Norwood, then in the county of Surrey.

Darwin was not invited to FitzRoy's funeral. Not surprisingly it was a small and very private affair arranged at short notice, although Sulivan attended (more through chance then design). The other *Beagle* officers were not given the opportunity to attend. Sulivan wrote to Darwin on 8 May 1865:

It was a very quiet and plain funeral, just what I think all funerals should be. Poor Mrs FitzRoy would go & the two daughters were with her. We all waited outside and walked after her carriage & the same back – the brothers only going to the house. It was a trying scene at the grave. Poor Mrs F and the girls looked dreadfully ill & Mrs F gave way very much. The coffin was plain black wood with 'Robert FitzRoy, born — died —' on a brass plate. You may suppose it was a trial for me, and the thought of old times and scenes that would be mixed up with it all.

Darwin enjoyed a longer life and would die with his wife and family members by his side on 19 April 1882. He reached the very respectable age of 73. He had earned and inherited considerable sums of money, including stocks and shares; also half of the estate of his brother Erasmus, who died in 1881. Darwin's personal wealth at his death was more than £250,000. His scientific colleagues persuaded his family that Westminster Abbey was the appropriate resting place. At his funeral Darwin's pallbearers included his staunchest supporters, Hooker, Huxley and Wallace.

After Darwin's return from the *Beagle* expedition he started to produce a steady stream of papers and publications that amounted over his lifetime to a substantial and significant body of research work. They included his *Journal of Researches* (1839), *The Structure and Distribution of Coral Reefs* (1842), *Geological Observations on South America* (1846), *A Monograph of the Fossil Sessile Cirripeds* (1854), *On the Origin of Species* (1859), *The Variation of Animals and Plants Under Domestication* (1868), *The Descent of Man and Selection in Relation to Sex* (1871), *The Expressions of the Emotions in Man and Animals* (1872), *The Power of Movement of Plants* (1880) and *The Formation of Vegetable Mould through the Action of Worms* (1881).

Darwin also edited parts of *The Zoology of the Voyage of H.M.S. Beagle*. This sumptuous work appeared in five parts (made up of nineteen numbers) between February 1838 and October 1843. He wrote some of the introductions and contributed other text. The main authors of the parts were Richard Owen (*Fossil Mammalia*), George Robert Waterhouse (*Mammalia*), John Gould (*Birds*), Leonard Jenyns (*Fish*) and Thomas Bell (*Reptiles*). The parts were lavishly illustrated, which greatly increased the production costs, although the publication as a whole benefited from a substantial government grant of £1,000. Many of the popular images relating to the *Beagle* expeditions derive from these volumes and they include a wide range of prehistoric skeletons, the Falkland fox, the land iguana, FitzRoy's dolphin *Delphinus Fitz-Royi* and the finch *Tanagra darwinii*.

But what of the fate of HMS *Beagle*, the ship that set all of Darwin's literary works in motion? Between 1837 and 1843 she completed a six-year expedition to survey large parts of the coast of Australia under the command of two of FitzRoy's former officers: Commander John Clements Wickham and assistant surveyor, Lieutenant John Lort Stokes, who would take over command when Wickham was invalided out of the ship in Australia.

The survey commenced with the western coast between the Swan River (modern-day Perth) and the Fitzroy River, Western Australia, then both shores of the Bass Strait at the south-east corner of the continent. In May 1839 the *Beagle* sailed north to survey the shores of the Arafura Sea opposite Timor. Wickham named the Beagle Gulf and Port Darwin, which was first sighted by Stokes, and which later gave its name to the city of Darwin, Australia.

After her return in 1843 *Beagle* would no longer sail as a naval vessel. In 1845 she was stripped out and converted as a static coastguard watch vessel and transferred to the Coast Guard Service to assist the Customs and Excise Office. She would be part of a network of floating lookouts to control smuggling, which was prevalent on the Essex coast and within the Thames estuary at that time. She was moored mid-river on the River Roach, which forms part of a maze of waterways in the marshes south of Burnham-on-Crouch. In 1850 oyster companies and traders petitioned for her to be removed as she was obstructing the river. She was moved to Pagelsham, on the east Essex coast between the rivers Crouch and Roach, and renamed *W.V.7*. (W.V standing for Watch Vessel).

In 1870 the *Beagle* was superfluous to requirements and she was sold for the paltry sum of £525 to Murray and Trainer, who, it has been assumed, were the scrap-dealers who broke her up. This may well have been the end of the story; however, investigations started in 2000 by a team led by Dr Robert Prescott of the University of St Andrews, resulted in their claim to have discovered the bottom of the *Beagle* lying in 18 feet of mud in the Essex marshes. If it is true, it is unlikely that she will now yield any further insights into her remarkable survey expeditions.

FitzRoy died a disappointed man. The psychological sledgehammer blow for him that almost certainly precipitated a spiral of depression was the unwelcome news he received in the summer of 1860. He had entertained hopes that his Fuegians might be of future assistance to British seafarers. In his mind they would remember the kindness bestowed upon them in England and help British sailors in need or distress. Tierra del Fuego would become a welcome British colonial stopover. But this proved a false hope. The news he received related to Jemmy Button's part (Jemmy was participating in another missionary project) in the massacre of Mr Phillips, Captain Fell, four seamen and two mates of the schooner *Allen Gardiner* by the natives in Woollya in Tierra del Fuego. It was no consolation to FitzRoy, but it was later found that the evidence against Jemmy Button did not entirely stack up. It was also discovered that York Minster was killed by a tribe in retaliation for a murder he had committed. No one was surprised to hear this news. Fuegia Basket probably died of old age. A letter from Arthur Mellersh to Darwin (25 January 1872) features reminiscences of the evening that he, Sulivan and

Wickham spent with Darwin at Down House, nearly a decade earlier. Among many topics, the men had discussed the mission to Tierra del Fuego and expressed some grave concerns about its relevance and future. Mellersh held out the hope that it would not 'improve' the people to extinction.

The following few words from the lengthy obituary published in *The Gentleman's Magazine* in June 1865 neatly sums up Vice-Admiral Robert FitzRoy's attitude to his life and career: 'His own life…was the price of his devotion to his duties'. His career certainly came first and his primary concern was to achieve something for public good.

Darwin knew that the *Beagle* expedition was the making of him. In a letter to FitzRoy on 20 February 1840, he wrote:

> However others may look back to the Beagle's voyage, now that the small disagreeable parts are well nigh forgotten, I think it far the most fortunate circumstance in my life that the chance afforded by your order of taking a naturalist fell on me – I often have the most vivid and delightful pictures of what I saw on board the Beagle pass before my eyes.– These recollections & what I learnt in Natural History I would not exchange for twice ten thousand a year.

Thanks to FitzRoy (albeit unwittingly) the legacy of the *Beagle* lives on. Through his pioneering weather work he tried to forewarn seamen, and the British public, of the conditions that lay ahead. Ironically he ignored the warnings from family and friends of the gathering 'storm clouds' that would ultimately destroy his life.

Opposite: Amblyrynchus Demarlii from *Reptiles* Part 5 No. 2 of *The Zoology of the Voyage of H.M.S. Beagle* by Thomas Bell. The land lizard shown here was smaller than the aquatic species. It ate cactus and the leaves of the acacia tree. Bell was a surgeon and zoologist. As President of the Linnean Society he chaired the meeting on 1 July 1858 at which Darwin and Alfred Russel Wallace first presented (read by others as they were not actually at the meeting) their theories on natural selection in a joint presentation of papers.

Below: Invitation to Charles Darwin's funeral. The *Pall Mall Gazette* thought him, 'the greatest Englishman since Newton', who was also buried in Westminster Abbey.

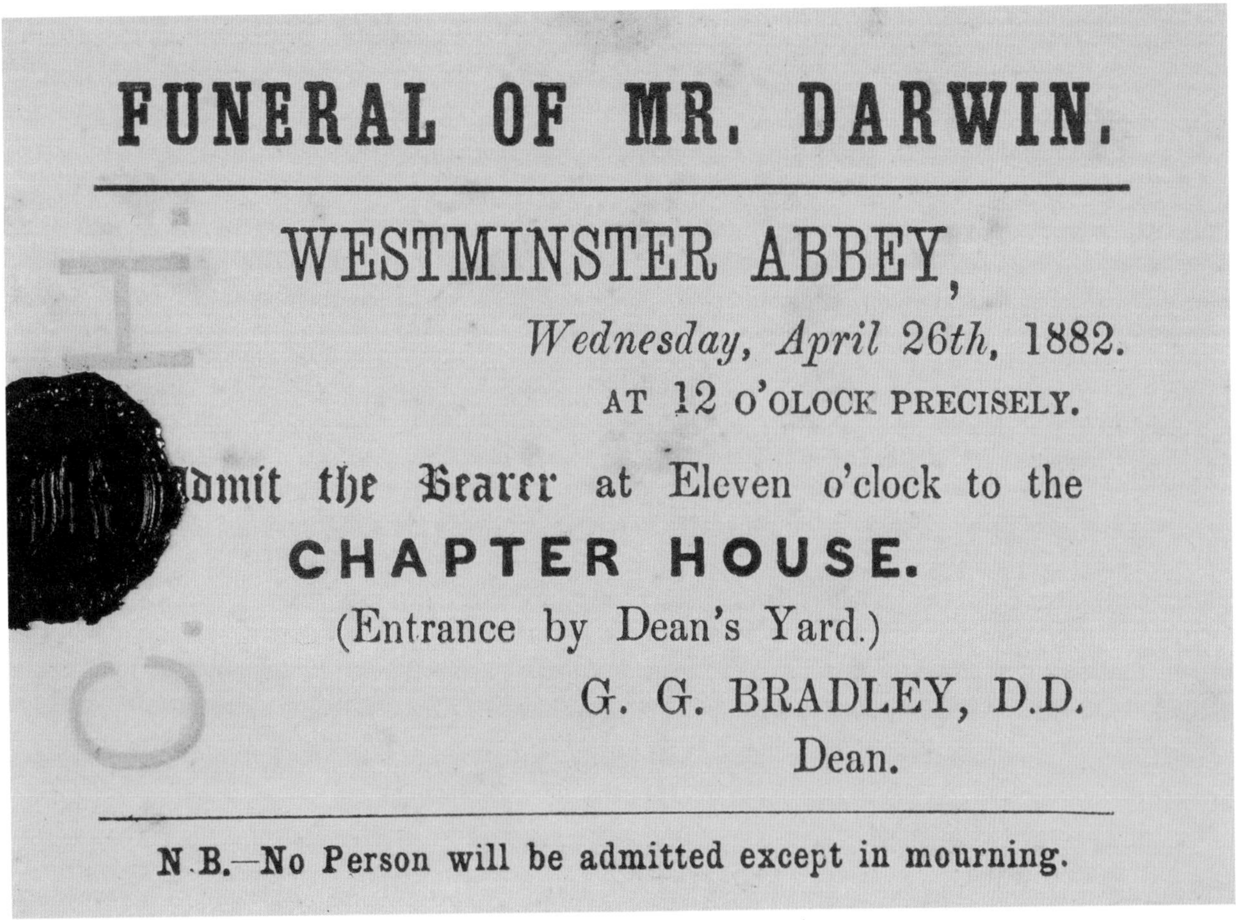

FUNERAL OF MR. DARWIN.

WESTMINSTER ABBEY,

Wednesday, April 26th, 1882.

AT 12 O'OLOCK PRECISELY.

dmit the Bearer at Eleven o'clock to the

CHAPTER HOUSE.

(Entrance by Dean's Yard.)

G. G. BRADLEY, D.D.
Dean.

N.B.—No Person will be admitted except in mourning.

Bibliography

Barlow, Nora, *Charles Darwin and the Voyage of the Beagle*, Pilot Press, 1945

Barlow, Nora, *The Autobiography of Charles Darwin*, Collins, 1958

Beer, Sir Gavin de, *Charles Darwin Evolution By Natural Selection*, Thomas Nelson and Sons, 1963

Bloxham, James, 'Martens' Mandate' (online extract from doctoral thesis relating to his work conserving Conrad Martens' *Beagle* sketchbooks I and III)

Bolton, G. C., 'John Lort Stokes (1812–1885)', *Australian Dictionary of Biography*, Vol. 2, Melbourne University Press, 1967, pp. 488-489

Bowlby, John, *Charles Darwin – A New Biography*, Hutchinson, 1990

Browne, Janet, *Darwin's Origin of Species: A Biography*, Atlantic Books, 2006

Browne, Janet, *Charles Darwin: Voyaging*, Volume I of a biography, Jonathan Cape, 1995

Browne, Janet, *Charles Darwin: The Power of Place*, Volume II of a biography, Jonathan Cape, 2002

Buscombe, Eve, *Artists in Early Australia and their portraits*, Sydney, NSW, 1978

Clowes, William Laird, *The Royal Navy: A History From the Earliest Times to the Present*, Sampson Low, Marston and Co., 1901

Cooke, Christopher, *FitzRoy's Facts and Failures*, Hall, 1867

Courtney, Nicholas, 'Gale Force 10: The Life and Legacy of Admiral Beaufort', *Review*, 2002

Covington, Syms, *The Journal of Syms Covington December 1831–September 1836* (see online version, Australian Science Archives Project)

Darling, Lois, 'The *Beagle*: A Search for a Lost Ship' [Natural History] 69(5):48-59 (1960)

Darling, Lois, 'H.M.S. *Beagle* Further Research or Twenty Years A-Beagling' (see *The Mariner's Mirror* Vol. 64, No. 4, November 1978 (or for the full article, *The Log of Mystic Seaport*, April 1972)

Darwin, Charles, 'The Voyage of the Beagle' *Journal of Researches Into The Natural History and Geology of the Countries visited during the voyage of HMS Beagle round the world, under the command of Captain Fitz Roy, RN*, John Murray, 2nd edition 1845

Darwin, Francis, *Life and Letters of Charles Darwin*, 3 vols, London: John Murray, 1887

Dawson, L. S., *Memoirs of Hydrography*, Cornmarket Press, 1969 (originally published in 2 vols)

Day, A., *The Admiralty Hydrographic Service: 1795–1919*, HMSO, 1967

Desmond, Adrian and Moore, James, *Darwin: The Life of a Tormented Evolutionist*, Warner Books, 1991

Dixson, William, 'Notes on Australian Artists Part II' (*Royal Australian Historical Society Journal and Proceedings* v.5, 1919, pp. 283-300)

Dundas, Douglas, *The Art of Conrad Martens*, Macmillan, 1979

Earle, Augustus, *A Narrative of a nine months' residence in New Zealand in 1827; together with a journal of a residence in Tristan D'Acunha, an island situated between South America and the Cape of Good Hope*, Longman, 1832

Earle, Augustus, *Sketches illustrative of the native inhabitants and islands of New Zealand, from original drawings*, New Zealand Association, 1838

Earle, Augustus, *The Wandering Artist: Augustus Earle's travels around the world, 1820–29*, Canberra, c.1994

Earle, Augustus, *Views in New South Wales, and Van Diemen's Land: Australian scrap book, 1830*, J. Cross, 1830

Ellis, Elizabeth, *Conrad Martens Life & Art*, State Library of New South Wales Press, 1994

Ellis, Frederick E., 'Robert FitzRoy Aboard HMS Thetis and HMS Ganges 1824–1828', see *The Mariner's Mirror*, Vol. 75, No.1, February 1989

Falk, Bernard, *The Royal Fitz Roys: Dukes of Grafton through four centuries*, Hutchinson, 1950

Fitzroy: letters to his family from HMS *Glendower, Hind, Thetis, Ganges* and *Beagle*, Cambridge University Library (Ref: Add.8853)

Ford, Colin, *Julia Margaret Cameron 19th Century Photographer of Genius*, National Portrait Gallery, London, 2003

Goodman, Jordan, *The Rattlesnake: A Voyage of Discovery To the Coral Sea*, Faber & Faber, 2005

Gribbin, John and Mary, 'FitzRoy: The Remarkable Story of Darwin's Captain and the Invention of the Weather Forecast', *Review*, 2003

Gruber, Jacob W., 'Who was the Beagle's Naturalist?', *British Journal for the History of Science* 4 (15):266-282 (1969)

Hackforth-Jones, Jocelyn, *Augustus Earle – Travel* (Artist Paintings and drawings in the Rex Nan Kivell Collection, National Library of Australia), Scolar Press, 1980

Hackforth-Jones, Jocelyn (and others), *Between Worlds Voyagers To Britain 1700–1850*, National Portrait Gallery, London, 2007

Hardie, Martin, *Water-Colour Painting in Britain*: I *The Eighteenth Century*; II *The Romantic Period*; III *The Victorian Period*, 3 vols, B. T. Batsford, 1967

Healey, Edna, *Emma Darwin The Inspirational Wife of a Genius*, Headline, 2001

Hordern, Marsden, *Mariners Are Warned! – John Lort Stokes and HMS Beagle in Australia 1837–1843*, Melbourne University Press, 1989

Howse, Derek (Ed.) *Background to Discovery: Pacific Exploration from Dampier to Cook*, Berkeley: University of California Press, 1990 (especially chapter VI by Sands, John O, 'The Sailor's Perspective: British Naval Topographic Artists')

Howgego, Raymond John, *Encyclopaedia of Exploration 1800 to 1850*, Hordern House, 2004

Huxley, Robert (Ed.), *The Great Naturalists*, Thames & Hudson, 2007

Joppien, Rudiger and Smith, Bernard, *The Art of Captain Cook's Voyages*, Yale University Press, 1985 and 1988 (3 volumes in 4 books)

Keevil, J. J., 'Benjamin Bynoe (1804–1865) Surgeon of H.M.S. Beagle', *Journal of the Royal Naval Medical Service* 35:251-268 (1949)

Keevil, J. J., 'Robert McCormick, R.N.', *Journal of the Royal Naval Medical Service* 29:36-62 (1943)

Kerr, Joan (Ed.), *The Dictionary of Australian Artists – Painters, Sketchers, Photographers and Engravers up to 1870*, OUP, Melbourne, 1992

Keynes, Richard Darwin (Ed.), *The Beagle Record* (selections from the original pictorial records and written accounts of the voyage of HMS *Beagle*), Cambridge University Press, 1979

Keynes, Richard Darwin, *Fossils, Finches and Fuegians: Darwin's Adventures and Discoveries on the Beagle*, Oxford University Press, 2003

King Papers, Mitchell Library, New South Wales (there is also a complementary collection of prints and drawings by several generations of the King family and others)

King, Philip Gidley, *Autobiography*, Mitchell library, New South Wales (unpublished manuscript)

King, Philip Parker, 'Some Observations upon the Geography of the Southern Extremity of South America, Tierra del Fuego and the Strait of Magalhaens'; *Journal of Royal Geographic Society of London* 1:155-175 (1832)

Lambert, Andrew, *The Last Sailing Battlefleet, Maintaining Naval Mastery 1815–1850*, Conway Maritime Press, 1991

Liebersohn, Harry, *Patrons, Travelers, and Scientific World Voyages 1750–1850* (a version of his paper presented to the *Seascapes* conference, Washington, D. C., 13–15 February 2003), American Historical Association

Lindsay, Lionel, *Conrad Martens: The Man and His Art*, Angus & Robertson, 2nd edition 1968.

Lloyd, Christopher, *The British Seaman 1200–1860: A Social Survey*, Collins, 1968

Lloyd, Christopher and Coulter, Jack L. S., *Medicine And The Navy 1200–1900*, Vol. IV 1815–1900 (especially Chapter V 'Surgeon-Naturalists')

Lucas, J. R., 'Wilberforce and Huxley: A Legendary Encounter', *The Historical Journal* 22 (2): 313-330, June 1979

McCormick, E. H. (Ed.), *Narrative of a Residence in New Zealand – Journal of a Residence in Tristan da Cunha by Augustus Earle*; Clarendon Press, Oxford (1966)

McCormick, Robert, *Voyages of Discovery in the Arctic and Antarctic Seas, and Round the World: Being Personal Narratives of Attempts to Reach the North and South Poles; and of an Open-Boat Expedition up the Wellington Channel in Search of Sir John Franklin and Her Majesty's Ships "Erebus" and "Terror," in Her Majesty's Boat "Forlorn Hope," under the Command of the Author*, Sampson, Low, Marston, Searle and Rivington, London; 2 volumes (1884)

Marquardt, Karl Heinz, *HMS Beagle: Survey Ship Extraordinary* (*Anatomy of the Ship* series), Conway Maritime Press, 1997

Martens, Conrad, *Journal of a voyage on board HMS Hyacinth, commenced May 19, 1833* (identified by Elizabeth Ellis, located in the Manuscript Department, State Library, New South Wales)

Mellersh, H. E. L., *FitzRoy of the Beagle*, Mason & Lipscomb, 1968

Moorehead, Alan, *Darwin and the Beagle*, Hamish Hamilton, London, 1969

Murray-Oliver, Anthony, *Augustus Earle in New Zealand*, Whitcombe & Tombs, 1968

Narrative of the Surveying Voyages of His Majesty's Ships Adventure and Beagle Between the Years 1826 and 1836: Describing Their Examination of the Southern Shores of South America, and the Beagle's Circumnavigation of the Globe, Henry Colburn, 1839: King, Phillip Parker (Vol. I), FitzRoy, Robert (Vol. II) and Darwin, Charles (Vol. III)

Neve, Michael and Messenger, Sharon, *Charles Darwin Autobiographies*, Penguin, 2002

Nichols, Peter, *Evolution's Captain*, Profile Books, 2003

O'Brian, Patrick, *Joseph Banks*, The Harvill Press, 1989

O'Byrne, William R. A., *Naval Biographical Dictionary*, John Murray, 1849

O'Grady, Frank, 'Philip Gidley King (1817–1904)', *Australian Dictionary of Biography*, Vol. 5, Melbourne University Press, 1974, pp. 29-30

O'Grady, Frank, 'Phillip Parker King (1791–1856)', *Australian Dictionary of Biography*, Vol. 2, Melbourne University Press, 1967, pp. 61-64

Quarm, Roger and Wilcox, *Masters of the Sea: British Marine Watercolours*, Phaidon Press, 1987

Ritchie, G. S., *The Admiralty Chart*, Hollis and Carter, London, 1967

Rodger, Nicholas, *The Wooden World; An Anatomy of the Georgian Navy*, Collins, 2nd edition 1986

Schetky, S. F. L., *Ninety Years of Work and Play: Sketches from the Public and Private Career of John Christian Schetky*, William Blackwood and Sons, 1877

Smith, Bernard, *European Vision and the South Pacific*, Yale University Press, 2nd edition 1988

Spencer, Harold Edwin, *Augustus Earle: a study of early nineteenth century travel-art and its place in English landscape and genre traditions*, Harvard University, 1967 (unpublished doctoral thesis on microfilm in British Library. Shelfmark MFR11574)

Stainton, Lindsay, *British Landscape Watercolours 1600–1860*, British Museum Publications, 1985

Stanbury, David, unpublished Notes and MSS relating to HMS *Beagle*, Charles Darwin, Christ's College, Cambridge Library

Stanbury, David, 'H.M.S. *Beagle*' (for a discussion of *Beagle*'s boats and sail arrangement) see *The Mariner's Mirror*, Vol. 65, No. 4, November 1979

Stanbury, David, *H.M.S. Beagle and the Peculiar Service* (Part I of *Charles Darwin A Centennial Commemorative 1809–1882*) edited by Roger G. Chapman and Cleveland T. Duval, Nova Pacifica, 1982

Stokes, John Lort, *Discoveries in Australia with an account of the Coasts and Rivers Explored and Surveyed During the Voyage of the H.M.S. Beagle in the Years 1837–1843*, T. and W. Boone, London, 2 volumes, 1846

Sulivan, Henry Norton, *Life and Letters of the Late Admiral Sir Bartholomew James Sulivan K.C.B. 1810–1890*, Murray, London, 1896

Taylor, James, *Marine Painting: Images of Sail, Sea and Shore*, Studio Editions, 1995

Thomson, Keith Stewart, *HMS Beagle: The Ship That Changed The Course Of History*, W. W. Norton & Co., 1995

Usborne, Julian, 'The Usborne Family Tree', www.usbornefamilytree.com

Vries-Evans, Susanna, *Conrad Martens: On the Beagle and in Australia*, Pandanus Press, 1993

Wilton, Andrew, *British Watercolours 1750–1850*, Phaidon, 1977

Wilton, Andrew and Lyles, Anne, *The Great Age of British Watercolours 1750–1850*, Prestel, 1993

Major repositories of birds, books, botanical specimens, fish, fossils, journals, letters, logbooks, manuscripts, navigational instruments, paintings, drawings, prints, photographs, scientific equipment, sculpture, dry and wet specimens relating to HMS *Beagle*, her officers, crew and supernumeraries can be found in the following public collections and major societies.

British Hydrographic Office, Taunton (Admiralty Charts, Pringle Stokes' *Journal* and FitzRoy's letters to Captain Francis Beaufort)

British Library, British Museum (Admiralty Charts)

Cambridge Zoology Museum (barnacles, beetles, finches, fish, fossils, and other wet and dry specimens)

Cambridge University Library (Darwin, FitzRoy letters, and those by officers and crew)

Christ's College, Cambridge (unpublished notes and manuscripts by David Stanbury; FitzRoy, Earle and Martens)

Down House, English Heritage (large range of personal Darwin items, including pictures and the original manuscript of Darwin's shipboard Diary)

John Murray Archive, Edinburgh (memory drawings by Philip Gidley King)

Linnean Society, London (portraits of Darwin)

National History Museum, London (specimens: fossils and finches, etc.)

National Library of Australia, Canberra (Earle and Martens pictures)

National Maritime Museum, Greenwich (Earle and Martens pictures; also Stokes collection)

National Portrait Gallery, London (portraits of Darwin, his family and associated scientists)

National Archives, Kew (Public Record Office) (*Beagle* logbooks, muster books, officers' journals and accounts and coastal profiles)

Oxford University Museum of Natural History (wet and dry specimens)

Royal Botanic Gardens, Kew (botanical specimens)

Royal Geographical Society, London (copies of Darwin and Fitzroy's certificates for election to Membership, volumes of the *Narrative*)

Science Museum, London (wet and dry specimens)

Sedgwick Museum of Earth Sciences, Cambridge (fossils)

State Library of New South Wales, Australia (Journal by Syms Covington and Martens pictures)

The Royal Society, London (FitzRoy's *The Weather Book*, Darwin and FitzRoy's certificates for election to Fellowship)

Picture Credits

Page 87: 'Breast Ploughing at Chiloe', engraving after Phillip Parker King from the *Narrative* (1839).
Page 190: 'Berkeley Sound Falklands Islands', after Conrad Martens from the *Narrative* (1839).
Page 192: 'Monte Video' (detail) after Conrad Martens from the *Narrative* (1839).

Index

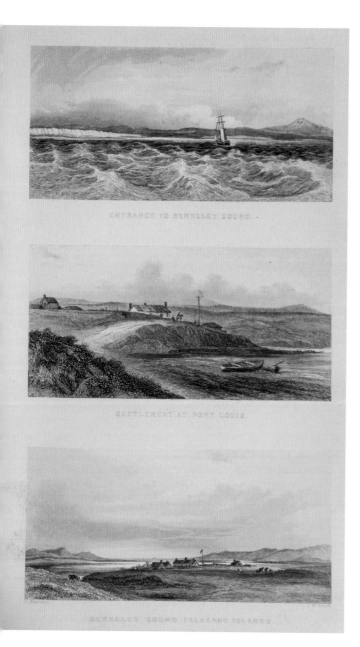

ENTRANCE TO BERKELEY SOUND.

SETTLEMENT AT PORT LOUIS.

BERKELEY SOUND FALKLAND ISLANDS.

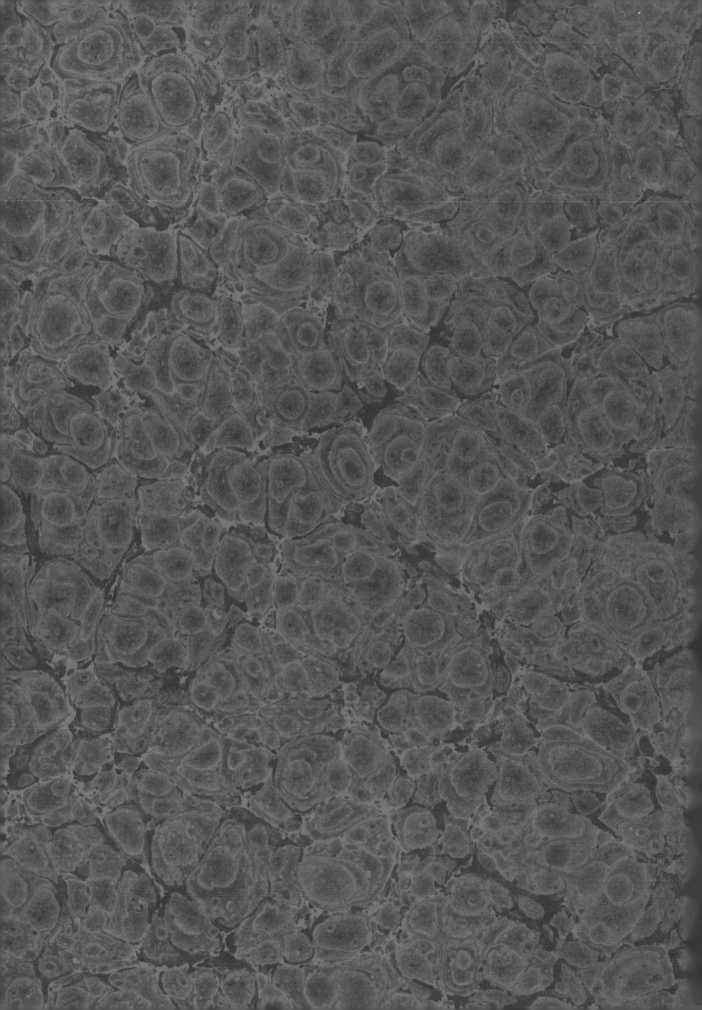